MUSIC OF THE UNITED STATES OF AMERICA

Richard Crawford, Editor-in-Chief
Jeffrey Magee, Executive Editor

---

1 Ruth Crawford: Music for Small Orchestra (1926); Suite No. 2 for Four Strings and Piano (1929)
Edited by Judith Tick and Wayne Schneider

2 Irving Berlin: Early Songs, 1907–1914
Edited by Charles Hamm

3 Amy Beach: Quartet for Strings (in One Movement), Opus 89
Edited by Adrienne Fried Block

White
N.Y.

Irving Berlin

Early Songs

I. 1907–1911

Edited by Charles Hamm

Recent Researches in American Music • Volume 20

Music of the United States of America • Volume 2, Part 1

Published for the
American Musicological Society
by

A-R Editions, Inc.

Madison

Published by A-R Editions, Inc.
801 Deming Way, Madison, Wisconsin 53717

Printed in the United States of America

Frontispiece: Irving Berlin in 1910. Courtesy of the Estate of Irving Berlin. Reproduced by permission.

Publication of this edition has been supported by a grant from the National Endowment for the Humanities, an independent federal agency.

The paper in this publication meets the minimum requirements of American National Standard for Information Sciences—Permanence of Paper for Printed Library Materials, ANSI Z39.48-1984.∞

Library of Congress Cataloging-in-Publication Data

Berlin, Irving, 1888–
[Songs. Selections]
Early songs / Irving Berlin ; edited by Charles Hamm.
1 score. — (Music of the United States of America ; 2)
For voice and piano.
Includes bibliographical references and index.
Contents: 1. 1907–1911.
ISBN 0-89579-305-9 (v. 1 : acid-free paper)
1. Popular music—United States—1901–1910. 2. Popular music—United States—1911–1920. I. Hamm, Charles. II. Title. III. Series.
M1630.18.B4E37 1994 94-20573
CIP
M

CONTENTS

FOREWORD

Music of the United States of America (MUSA), a national series of scholarly editions, was established by the American Musicological Society (AMS) in 1988. In a world where many developed nations have gathered their proudest musical achievements in published scholarly form, the United States has been conspicuous by its lack of a national series. Now, with the help of collaborators, the AMS presents a series that seeks to reflect the character and shape of American music making.

MUSA, planned to encompass forty volumes, is designed and overseen by the AMS Committee on the Publication of American Music (COPAM), an arm of the society's Publication's Committee. The criteria foremost in determining its contents have been: (1) that the series as a whole reflect breadth and balance among eras, genres, composers, and performance media; (2) that it avoid music already available through other channels, duplicating only where new editions of available music seem essential; and (3) that works in the series be representative, chosen to reflect particular excellence or to represent notable achievements in this country's highly varied music history.

The American Musicological Society's collaborators in the national effort that has brought MUSA to fruition include the National Endowment for the Humanities in Washington, D.C., which has funded MUSA from its inception; Brown University's Music Department in Providence, Rhode Island, which provided the project's original headquarters; the University of Michigan School of Music, where MUSA now makes its home; A-R Editions, Inc., the publisher, on behalf of AMS, of the MUSA series; and the Sonneck Society for American Music, which, through its representative to COPAM, has provided advice on the contents of MUSA.

Richard Crawford, Editor-in-Chief

PREFACE

This edition presents the 190 copyrighted songs published between 1907 and 1914 for which Irving Berlin wrote lyrics, music, or both. It also includes an appendix whose contents document more fully Berlin's musical activity in those years: two of his songs copyrighted but never published; transcriptions of four songs recorded but never published; nine unpublished songs that were never copyrighted; one extended strophic poem, with a final verse intended to be sung; three pieces written or arranged for piano by Berlin; three piano rags by other composers later made by Berlin into ragtime songs; and one song by Berlin in a piano arrangement by another musician.

The period 1907–14 represents the first phase of Berlin's songwriting career, when his primary concern was with individual songs. His first published song dates from 1907, and subsequent songs share similar stylistic traits until late 1914, when his attention turned to complete shows. The songs from the show *Watch Your Step*, which opened at the Globe Theatre on 8 December 1914, are not included in the present edition because they belong to the second phase of his career.

Berlin wrote many other songs during these early years. The Irving Berlin Collection in the Library of Congress (LC-IBC), for example, contains handwritten or typed lyrics of more than fifty additional unpublished songs from the period, and Berlin's working inventory of songs written between 1912 and 1914 lists another twenty-five unpublished and unrecovered ones.[1]

---

[1]Other songs attributed to Berlin are excluded from this edition for lack of firm evidence of his authorship, such as an attribution to him in the published sheet music, his name as author in copyright records, or other documentary evidence. Among them are:

"The Belle of the Barber's Ball." In 1911 Berlin and George M. Cohan coauthored a song of this title that was never published or copyrighted and cannot be located. J. Ord Hume's band arrangement of the piece, incorporating Berlin's "Yiddle, on Your Fiddle, Play Some Ragtime" in the introduction, was published in *Boosey's March Journal* (no. 233), and copyrighted in the United States on 7 February 1912.

"Brand New." Sheet music covers for several songs from the show *She Knows Better Now*, including "Ragtime Mocking Bird" and "I Want to Be in Dixie," list this song as being by Berlin and Ted Snyder, as does a catalog issued by the Irving Berlin Music Corporation in 1948, but copyright files and deposit cards in the Library of Congress list Cecil Mack and Chris Smith as authors.

"Lonely Moon," "My Heather Bell," and "Take Me Back to the Garden of Love." The 1948 catalog of the Irving Berlin Music Corporation lists Berlin as the coauthor of these songs, with E. Ray Goetz (his brother-in-law) and A. Baldwin Sloane. But neither the published sheet music nor the copyright deposit cards in the Library of Congress mentions Berlin. In addition, the titles aren't included in a list of songs drawn up by Berlin in 1916, nor are the pieces found in the six volumes entitled "My Songs" brought together and bound the same year. A later catalog, *The Songs of Irving Berlin*, published by the Irving Berlin Music Corporation in 1957, eliminates the songs and corrects many other mistakes in the earlier catalog. Berlin may have had a hand in writing them and then declined to take credit, but there is no documentation of that.

Despite Berlin's stature as the era's leading songwriter, the majority of the songs in this edition are difficult to obtain. Copies of the original sheet music can be found only in research libraries or private collections.[2] When Berlin established his own publishing company in the early 1920s, he reissued most of his early songs, some from the original plates and some in new arrangements, but these have become almost as rare as the original sheets. From time to time Berlin's publishing company brought out folio collections of his songs arranged for keyboard or in simplified settings for voice and piano, but these volumes contain at best a handful of the songs from 1907–14, none in their original form, and they also have become rare items. Berlin never gave permission for his music to be brought out by other publishers, thus none of his songs was included in anthologies of Tin Pan Alley songs published before copyright protection began expiring on his earliest songs. A few of these pieces may be found in more recent anthologies, and many more are contained in an ambitious but carelessly done multivolume facsimile collection. (Anthologies containing Berlin's early songs are listed in the critical commentary for this edition.) Most recently, the Irving Berlin Music Company has published six folios of songs spanning his entire career, including a dozen from the period covered by the present edition. In all, though, no more than a third of the songs written between 1907 and 1914 are presently available in easily obtainable publications, and not one in a critical edition.

Many of Berlin's early songs rank among the most popular and influential music of the period. Some are no more than serviceable, however, while others are forgettable and forgotten. Why not, then, offer an anthology of Berlin's best songs rather than all of them?

MUSA is intended to serve as a resource for scholars and students, and to furnish performers with repertories of American music not otherwise available, or available only in corrupt form. An anthology, by definition, would include some of these songs while excluding others, the selection determined by the editor's judgment of the quality and historical importance of each song or by mere personal preference. I have my own opinions about these songs, individually and collectively, but it's not the function of a scholarly edition to impose the judgment and taste of the editor on the user, even through so subtle a device as withholding some of the repertory. Moreover, offering all of the songs, in chronological order, gives a far more accurate and complete picture of Berlin's development as a songwriter than could emerge from an anthology.

"Rum Tum Tiddle." This copyrighted and published song by Jean Schwartz and William Jerome was arranged for band by J. Ord Hume for *Boosey's March Journal* (no. 252) with an introduction incorporating Berlin's "Stop, Stop, Stop (Come Over and Love Me Some More)." It was copyrighted in the United States on 30 April 1913.

"Too Many Eyes Are Making Eyes at Me" and "The Yiddisha Rag." These songs were attributed to Joseph H. McKeon, W. Raymond Harris, and Harry M. Piano, in the sheet music and in the Library of Congress copyright files. Vince Motto, among others, suspects that *Harry M. Piano* was a nom de plume for Berlin and thus suggests that he was a coauthor; but a letter from Harris to James Fuld, cited in the critical commentary for "Oh, What I Know About You," makes it clear that Harry Piano was a real person. Even though Berlin had a hand in the lyrics of the latter song, there is no evidence of his involvement with "Too Many Eyes" or "The Yiddisha Rag."

[2]James J. Fuld of New York City, a private collector, has brought together a virtually complete set of the first-edition sheet music of these songs. The Lester S. Levy Collection of Sheet Music, almost as complete, is now housed in the Milton S. Eisenhower Library of Johns Hopkins University in Baltimore. The Library of Congress in Washington, D.C., and the British Library in London, as copyright repositories for the United States and the British Commonwealth respectively, should have deposit copies of all of these songs, but vagaries of cataloguing and storage make it difficult and sometimes impossible to locate one song or another. The New York Public Library has put a large number of Berlin's early songs onto microfilm. Various other public and university libraries have acquired large sheet music collections, but most of these, including the Starr Collection at Indiana University and the Driscoll Collection in Chicago's Newberry Library, are catalogued only partially or not at all. The John Hay Library at Brown University has one of the best medium-sized collections, and smaller ones at Dartmouth College and the University of Virginia are particularly useful because their contents have been entered into computerized data bases.

Most of the songs in this edition have both lyrics and music attributed to Berlin, but those written in collaboration with others are also included. The making of a Tin Pan Alley song was often a corporate process, with two or more songwriters working together on lyrics and music. Copyright credit was often assigned with royalty distribution in mind, and the fact that dozens of songs were published with lyrics attributed to Berlin and music to Ted Snyder doesn't mean that Berlin had no hand in shaping the music, or that they should be excluded from the canon of his pieces. Berlin's contribution to a given song cannot be untangled from that of his collaborator(s). To attempt to do so, in fact, would give a distorted image of what actually went on in the studios of Tin Pan Alley publishing houses.

Berlin himself understood songwriting as a collaborative process. In 1916 he drew up a list of songs for which he claimed authorship and had copies of each piece bound into six volumes with "My Songs" stamped on the covers. Included are all those he wrote in collaboration with others as well as songs attributed solely to him. (This list is now in LC-IBC and the six volumes are in the possession of Berlin's daughters.)

Both composition and production involved collaboration. Berlin dictated his melodies to a staff musician, who wrote them down as lead sheets and then worked with the songwriter on harmonization and a piano accompaniment. After further polishing, the song was sent off to an engraver, where one or more craftsmen prepared the plates from which the song would be printed. With so many people involved in shaping the final product, and with frequent changes of personnel among staff musicians and engravers, details of musical notation, orthography, and layout vary from song to song. Different indications are used to indicate the da capo or dal segno return after the chorus; cautionary accidentals and articulation marks appear inconsistently; extender lines may take the form of dotted lines or extended dashes, or may be missing altogether; there is often incorrect division of words into syllables or no division at all; the sense of the text is sometimes confused by the absence or misplacement of punctuation; and capitalization of song titles is erratic.

Since these songs were engraved with a certain amount of elegance in a large format, furnishing the singer and accompanist with easy-to-read text and music, a facsimile edition would have had much to recommend it. But innumerable changes in the original plates would have been necessary to eliminate the imperfections and inconsistencies just noted. So the songs are newly set here, to provide maximum consistency in the notation of text and music and a common visual style.

Some European writers have criticized the "notational centricity" that musicology brings to the study of popular music, arguing that a focus on written scores tends to obscure the ways in which popular music is "rearticulated" in performance through pitch inflection, rhythmic fluctuation, manipulation of timbre, and even reshaping of overall structure.[3] If the traditions of popular music encourage performers to shape their own versions, then, what is the point of an edition of Berlin's songs based on published piano-vocal arrangements, with care taken to get every detail just right?

Popular music of the Tin Pan Alley era combined the traditions of notated and oral music. Songwriters created their pieces by ear; many of them, including Berlin, couldn't write their music down accurately. But eventually they worked their lyrics and music into a precise shape that seemed right to them. This version of the song was then "frozen" in musical notation by a musician skilled in transcription and published in a piano-vocal arrangement. Performance took place in one of two environments: (1) in home settings by amateurs, for solitary pleasure or in a circle of family and friends; many such performances would be note-for-note readings of the sheet music, for voice

[3]See Philip Tagg, "Analysing Popular Music: Theory, Method and Practice," *Popular Music* 2 (1982): 37–68, and Richard Middleton, *Studying Popular Music* (Philadelphia: Open University Press, 1990), particularly 104–6.

and piano or for piano alone (since the right hand of the accompaniment almost always doubles the voice); (2) on stage or in recording studios by professional singers. According to performance conventions of the day, singers could embellish or change the tune, or even deliver it in a semi-spoken fashion; tempo markings were taken as general suggestions only, as recordings of the period attest; the various sections of a song (introduction, vamp, verses, chorus) could be arranged in different sequences; and the piece could be transposed to any key suitable for the singer. Thus a professional performance might differ considerably from what is found in the published sheet music. But as I have written elsewhere:

> This transformation could take place only after the song itself had been created. Just as Bach or Beethoven couldn't fashion variations on a theme until there was such a theme, so performers [of popular music] couldn't create their own versions [of songs] . . . until they were written, and we can't understand and appreciate these versions until we know what they started from.[4]

In one way or another, then, the published piano-vocal scores of Tin Pan Alley songs provided the basis for contemporaneous performances of these pieces, just as the present edition can serve as the basis for any type of performance today.

[4]Charles Hamm, review of *The Music of Stephen C. Foster: A Critical Edition*, ed. by Steven Saunders and Deane L. Root, in *Journal of the American Musicological Society* 45 (Fall 1992): 526.

ACKNOWLEDGMENTS

Many people have contributed in one way or another to this edition during the almost four years it has taken to bring it to completion.

Richard Crawford encouraged and guided me every step of the way, from the moment the idea of a complete, critical edition of Berlin's early songs popped into my head during a conversation with him in Cambridge in October 1989, right through the final stages. His remarkable knowledge of the entire field of American music and its literature, his unselfishness in sharing this knowledge, his formidable and uncompromising editorial skills, his unflagging optimism and common sense—all these facilitated and strengthened every aspect of this edition in ways that only he and I can ever know.

Christopher Hill, Publications Manager of A-R Editions, was likewise involved from the early stages to the end, offering sound advice on the editing of popular music, counseling me on various artistic and technical matters, and facilitating the production process.

Among the many people contributing to the job of identifying, locating, and obtaining copies of all these songs were Bonny Wallin, Special Collections Assistant for Theatre and Music at Dartmouth College's Baker Library; Rosemary Cullen, Curator of the Harris Collection of the John Hay Library at Brown University; Cynthia Requardt, Kurrelmeyer Curator of Special Collections of the Milton S. Eisenhower Library of Johns Hopkins University, and Joan Grattan, Manuscripts Assistant of the collection; a host of people at the Music Division of the Library of Congress, including James Pruett, Gillian Anderson, Wayne Shirley, Ray White, and Mark Horowitz; Malcolm Turner and members of his staff at the British Museum in London; and Joseph Marchi of the Center for the American Musical at Canada College.

Private sheet music collectors were unfailingly generous in sharing information about and xerox copies of their treasures. I'm particularly indebted to Vincent Motto, whose *The Irving Berlin Catalog,* published as volumes 6, no. 5, and 8, no. 1, of the *Sheet Music Exchange,* is a rich source of information about Berlin's songs, and who sent xerox copies of many items that I couldn't locate elsewhere. After I realized the importance of collating period sound recordings with the printed first editions, I experienced the same cooperation and generosity from Paul Charosh and Eric Bernhoft, who between them furnished me with cassette dubbings of some fifty early recordings of these songs. And once I yielded to the inevitable and began my own sheet music collection, Ralph and Marilyn Andreson and Paul Riseman helped with that.

Wayne Schneider, the first Executive Editor of MUSA, was enthusiastically involved in every aspect of this edition from the beginning. He helped locate songs and other material; read preliminary sketches and various drafts of the introduction, editorial method, and critical notes; played and sang through each of these songs in a note-by-note check of my preliminary editing; discussed issues of editorial procedure when they were still fluid; and offered useful, penetrating comments on the musical style of these songs, both individually and collectively.

Jeff Magee, his successor, became involved in the project in the critical final stages. He reviewed every aspect of the edition with a fresh eye, raising perceptive questions that contributed to a tightening and reorganization of the introductory essay, editorial method, and critical notes.

In later stages, Wayne Shirley and Allan Atlas gave expert critical reviews of the entire edition, particularly the introductory essay, offering fresh and expert insights and suggestions.

After work on the edition was well underway, papers from the Irving Berlin estate, containing material of critical importance, were deposited in the Library of Congress. Even though the collection was uncatalogued and not yet generally available, the generous intervention of Berlin's three daughters, Mary Ellin Barrett, Linda Emmet, and Elizabeth Peters, and of Alton Peters, enabled me to consult and use this material. The Berlin family also arranged for relevant materials in the files of the Irving Berlin Music Company to be made available to me. As a result of their help, the inclusion of new and unpublished material in the edition makes it possible to see Berlin as the thoroughly professional songwriter he was, revising and reshaping his songs and not releasing them for publication until they were just the way he wanted them.

Theodore S. Chapin, Executive Director of the Rodgers and Hammerstein Organization (which now owns the catalogue of the Irving Berlin Music Company), was always cooperative, as were members of his staff, particularly Nicole Gillette, Kurt Reighley, and the late Hilda Schneider. James J. Fuld allowed me to examine his unmatched collection of first editions of the Berlin songs and offered excellent advice on several aspects of the edition, based on his rich experience as a collector and scholar of sheet music.

Mary Ellin Barrett, who was writing a biography of her father while this edition was in progress, encouraged, supported, and facilitated my work on numerous occasions. Michael Walsh contributed in a way that only he could have, at a critical point. And last, but anything but least, Robert Kimball, who is editing the complete lyrics of Berlin in collaboration with Linda Emmet, was helpful in many extremely important ways: he called my attention to discrepancies between his and my data on the early songs, offered wise counsel on various aspects of the edition, and, because of his great affection for the Berlin family and his love of the music, acted as an intermediary in dealing with the complex issues of permission and copyright for unpublished materials, drawing on his legal training and wide experience in writing and editing.

IRVING BERLIN AND EARLY TIN PAN ALLEY

Historians of American popular music have long since agreed that a period marked by substantial homogeneity of musical and textual style, and of production and marketing techniques, stretched from the last decade of the nineteenth century into the 1950s. *Tin Pan Alley* was a term allegedly coined in 1903 by the songwriter Monroe Rosenfeld for New York's Twenty-eighth Street, where the offices of many major music publishers were then clustered. It was soon adopted within the trade and by journalists as a label for the entire popular song industry, even after most companies shifted their operations further uptown. Isaac Goldberg expanded its usage to include the music itself,[1] and for the past half century *Tin Pan Alley* has had three separate but interlocking meanings: the historical period in question, the music industry of the time, and the style of song produced by this industry.

To deal most effectively with this era and Irving Berlin's place in its history, one must examine not only the primary texts, the songs themselves, but also such social factors as the dichotomy between the two performance venues for popular music—the "home circle" and the popular theater—and the critical role played by the culture of New York City in shaping Tin Pan Alley song.

The Tin Pan Alley Era and Its Antecedents

Arthur M. Schlesinger, Jr., writes:

> Having cleared most of North America of their French, Spanish, and Dutch rivals, the British were free to set the mold. The language of their new nation, its laws, its institutions, its political ideas, its literature, its customs, its precepts, its prayers, primarily derived from Britain. . . . The smelting pot thus had, unmistakably, an Anglocentric flavor. For better or worse, the white Anglo-Saxon Protestant tradition was for two centuries—and in crucial respects still is—the dominant influence on American culture and society.[2]

The formative years of Tin Pan Alley overlapped with the last decades of the Victorian era. It's no mere happenstance that this historical period bears the name of a British monarch, in the United States as well as in Britain; the era was marked in this country not only by the continuing dominance of Anglo-Saxon Protestant culture but also by

[1]Isaac Goldberg, *Tin Pan Alley: A Chronicle of the American Popular Music Racket* (New York: John Day, 1930).

[2]Arthur M. Schlesinger, Jr., *The Disuniting of America: Reflections on a Multicultural Society* (New York: Norton, 1992), 27–28.

a reliance upon England for social and cultural models that might help deal with the new tensions and contradictions of the industrial age. As Daniel Walker Howe puts it:

> Victorian culture had a class derivation, as well as an ethno-religious one. It was bourgeois in origin, and the era of its flourishing coincides with that of the predominance of the bourgeoisie in Western civilization. The class within which Victorian culture took shape was largely a product of the industrial revolution—indeed, in some ways it was not much older than the industrial proletariat.[3]

According to Howe, the "most active and articulate propagators" of American Victorian culture were clustered in the urban Northeast, "mostly middle-income, mostly Whig-Republican, literary men and women" who wanted to "humanize the emergent industrial-capitalist order by infusing it with a measure of social responsibility, strict personal morality, and respect for cultural standards."[4] To accomplish this, they promoted a value system whereby people were "to work hard, to postpone gratification, to repress themselves sexually, to 'improve' themselves, to be sober [and] conscientious."[5]

These Victorians assumed that newcomers from abroad, as well as Americans in other sections of the country, should learn to conform to their vision. Henry James, returning home in 1904 after some years abroad, noted with approval the process whereby a constant flood of immigrants was being transformed, in a single generation, into "Americans":

> The machinery is colossal—nothing is more characteristic of the country than the development of this machinery, in the form of the political and social habit, the common school and the newspaper; so that there are always millions of little transformed strangers growing up in regard to whom the idea of intimacy of relation may be as freely cherished as you like.[6]

Public schools, as critical cogs in this machinery, were expected to instill values in their students consistent with the American version of the Victorian ethic: pride in hard work, faith in God and eternal life, cheerful obedience to authority, the sanctity of family, the pleasures of comradeship, and appreciation of the beauties of nature and the fine arts. Pupils were exposed to these concepts through direct instruction and also through exposure to appropriate pieces of literature, music, and art. Musical literacy was considered a highly desirable skill, and entire generations of children learned musical notation as part of their general elementary and secondary education. Graded instruction books and anthologies of solo and part-songs served as learning texts, and simplified versions of popular songs, chosen from those with lyrics and music consonant with Victorian morality, were a major component of these collections. And for those who mastered some skill in singing or playing while in school and wanted to continue musical activity afterwards, in their family or social circle, composers and music publishers turned out a flood of vocal and piano pieces for adult amateurs pitched at the same didactic level as the music used in schools.

Beautiful Melodies, an anthology of such music brought out in 1895, is subtitled *A Choice Collection of Popular Music Comprising Songs for the Fireside, Social Gatherings, Concerts, Etc. Etc.* The preface is nothing less than a Victorian manifesto:

> Emerson says beauty is its own excuse for being—having a right to exist for the simple reason that it is beauty. The same may be said of a choice collection of Music. Like the morning light, it needs no apology for going forth, cheering the heart, entertaining the social circle, thrilling the public assembly, and charming multitudes with "the concord of sweet sounds."
>
> There is something [here] for every mood of mind and heart—for the joy that clamors for expression in melody, and the sorrow that is soothed by the mysterious influence of music. The grave, the pathetic, the cheerful and buoyant, the romantic, the patriotic, the devout—all these

---

[3]Daniel Walker Howe, "Victorian Culture in America," in *Victorian America,* ed. Daniel Walker Howe (Philadelphia: University of Pennsylvania Press, 1976), 10.

[4]Ibid., 12.

[5]Ibid., 17.

[6]Henry James, *The American Scene* (1907; reprint, Bloomington: Indiana University Press, 1968), 120.

various emotions and sentiments here have voice and expression. . . . It has been said that "Music is the speech of angels": it is no less true that these sublime harmonies are the natural language of the human heart.

In compiling this volume the author has been careful to introduce no songs of an objectionable character, as much care having been bestowed in the selection of words as music. The book is, therefore, suitable for every family; without any reserve whatever it can be placed in the hands of the young without fear of evil suggestion or any injurious effect.[7]

The collection includes songs, originally published as individual items of sheet music and thoroughly Victorian in substance and style, by such nineteenth-century British and American songwriters as Henry Bishop, J. L. Molloy, H. P. Danks, Virginia Gabriel, Claribel (Charlotte Alington Barnard), Henry Tucker, and Septimus Winner. Significantly absent are the "people's songs" of Stephen Foster and George F. Root, which belong to another tradition.[8]

Gathered Pearls, a similar collection of music and poetry "comprising a vast treasury of all that is most captivating, most soul-stirring, most pathetic, most sublime, most lofty in thought, glowing in description and eloquent in language expressly designed to be a companion in the home, a source of entertainment and instruction in the household," pronounces itself "so elevated in tone and so pure in sentiment that it can be welcomed into every family without fear that any corrupting influence will go with it."[9] Yet another anthology of this sort, *Perfect Jewels,* is subtitled *The Music of Home, Country and Heaven.*[10]

Gilbert Chase, in his seminal book on American music, suggests that music of this sort was part of the "genteel tradition"—a term borrowed from George Santayana—and was thus "characterized by the cult of the fashionable, the worship of the conventional, the emulation of the elegant, the cultivation of the trite and artificial, the indulgence of sentimentality, and the predominance of superficiality."[11] These pieces emulate the style of contemporary art songs. Melody and harmony are triadic and tonal, spiced with simple modulations, mild chromaticism, and expressive nonharmonic tones; melodic and harmonic sequence is common; climaxes often occur shortly before the final cadence; accompaniments may require more than elementary pianistic skill; and song forms are patterned after those favored by Schubert, Abt, and Brahms. In fact, the general style of genteel songs is so compatible with that of classical music of the day, albeit somewhat simplified, that these pieces appear side by side with songs by Schubert, Mozart, Mendelssohn, Bellini, Spohr, and their peers with no stylistic clash. It was this very pretension to high art that made them so appropriate for didactic purposes.

The popular songs "of Heaven, Country, and Home," then, played a critical role in the hegemonic reinforcement of Victorian ideology in the nation's schools and home circles.

The genteel repertory was not the only popular music of the nineteenth century, however, nor were the country's schools and home circles the only venues for performance of popular songs. American songwriters wrote for the popular musical stage as well, producing songs that differed in style from those described above, expressed different sentiments, and targeted a different audience.

Early in the nineteenth century, American theaters offered mixed entertainment to mixed audiences. A single theater—New York's famous Park Theatre, for instance—

---

[7]*Beautiful Melodies* (Philadelphia and Chicago: J. H. Moore, 1895), i–ii.

[8]For an extended profile of Root as a songwriter "in tune with society, sharing its tastes and working cheerfully and effectively within the musical life of a democracy," see Richard Crawford, *The American Musical Landscape* (Berkeley and Los Angeles: University of California Press, 1993), 151–83.

[9]*Gathered Pearls* (Chicago: Smith & Simon, 1895), iii–iv.

[10]*Perfect Jewels* (Boston: M. R. Gately, 1885).

[11]Gilbert Chase, *America's Music: From the Pilgrims to the Present* (New York: McGraw-Hill, 1955), 165. In the revised third edition (Urbana: University of Illinois Press, 1987), Chase dropped all reference to the "genteel tradition."

could be the site for the performance of comic operas, serious operas, legitimate dramas, farces, melodramas, dramatic or humorous recitations, classical or popular vocal music (or a mixture of both), instrumental music, and miscellaneous other entertainments. A single evening's program might be a bewildering mixture, at least by present-day standards, of highbrow and lowbrow events.[12] Audiences were similarly heterogeneous, with theaters designed to accommodate different classes. The urban elite occupied the three tiers of box seats, joined by "wealthy visitors to their cities: country merchants and gentry farmers, sea captains and merchants from other ports and a variety of other well-to-do travellers." But others were in attendance as well:

> Male apprentices, journeymen, servants, and African-Americans generally congregated in the gallery, the rear balcony containing the lowest-priced seats. The "middling classes," to use the designation of a journalist in 1825, usually sat on wooden benches in the pit, that area of the theatre below the boxes roughly corresponding to the present-day orchestra. The pit also held some of the better-off workers.[13]

Lawrence Levine argues that the separation of high and low culture in America dates from after 1850, but it had in fact begun well before midcentury,[14] though this development is somewhat obscured by the appropriation of European art music by proponents of religiously-based social reform.[15] The Boston Handel and Haydn Society was founded in 1815 for the purpose of "introducing into more general use the works of Handel and Haydn and other eminent composers" and "improving the style of performing sacred music,"[16] and The Musical Fund Society of Philadelphia was formed in 1820 expressly for the performance of classical music, as were similar organizations in other cities. The Astor Place Opera House (1847) was built for performances of "serious" opera only; the New York Academy of Music opened in 1854 "for the purpose of cultivating a taste for music by concerts and operas"; and the Philadelphia Academy of Music opened in 1857 with a similar mission. These organizations and performance spaces resulted from a conscious attempt by the urban elite to "separate high culture from low and to purify artistic experiences so as to achieve aesthetic and spiritual elevation rather than mere entertainment."[17]

Concurrently, popular culture was finding its own audiences and its own venues. Some theaters that had once offered a mixture of popular and elite cultural events, the Bowery and Chatham theaters in New York for instance, by midcentury offered chiefly "lowbrow" attractions such as minstrel shows, in a calculated appeal to working-class audiences. Bars, beer halls, and other performance spaces strictly for popular entertainment proliferated, and in 1865 Tony Pastor's Opera House in lower Manhattan opened as a showcase for popular entertainers. Though members of the elite classes might sometimes attend the popular theater, the audience was made up mostly of socially

[12]These terms are defined and discussed in their historical context by Lawrence Levine in *Highbrow/Lowbrow: The Emergence of Cultural Hierarchy in America* (Cambridge: Harvard University Press, 1988).

[13]Bruce A. McConachie, *Melodramatic Formations: American Theatre and Society, 1820–1870* (Iowa City: University of Iowa Press, 1992), 7–8.

[14]See Paul Charosh, " 'Popular' and 'Classical' in the Mid-Nineteenth Century," *American Music* 10 (Summer 1992): 117–35, and Charles Hamm, "The Dawning of Classical Music in America (1825–65)," in *Music in the New World* (New York: Norton, 1983), 195–229.

[15]For a discussion of the use of European music by an early American reform movement, see Richard Crawford, " 'Ancient Music' and the Europeanizing of American Psalmody, 1800–1810," in *A Celebration of American Music: Words and Music in Honor of H. Wiley Hitchcock*, ed. Richard Crawford, R. Allen Lott, and Carol J. Oja (Ann Arbor: University of Michigan Press, 1990), 225–55.

[16]Charles C. Perkins and John S. Dwight, *History of the Handel and Haydn Society of Boston, Massachusetts*, vol. 1 (Boston: Alfred Mudge & Son, 1883), 39.

[17]McConachie, *Melodramatic Formations*, 235.

marginalized people: the working classes, the urban poor, blacks, immigrants, and first-generation Americans.[18]

This physical separation of elite and popular theater in the middle decades of the nineteenth century was inevitable, given growing class tensions in the country and the fact that the theater itself became one focus of social discontent. "The Farren riot of 1834 and the Astor Place riot of 1849 began as theatre riots but rippled outward to encompass the racial or class tensions of New York City," according to Bruce A. McConachie. "Rioters at the Astor Place Opera House were avenging the treatment of their democratic hero, Edwin Forrest, at the hands of William Macready, a symbol, to them, of aristocratic villainy."[19]

As popular theater split away from elite theater, it began to offer an alternative to, and often subtle criticism of, America's white, British-based, Protestant culture. And it soon came under attack by those determined to perpetuate the latter heritage: clergy, educators, intellectuals, and other public figures. Some of this criticism seems warranted, since rowdy behavior and assorted vices were common. *The Spirit of the Times,* for example, reported the scene at the bar of the Bowery Theatre in 1847:

> [The patrons were] drinking, swearing, smoking, chewing tobacco, knocking each others' hats down over the eyes, and in a thousand such delicate and animal spirits, stimulated and excited to a sort of good-natured madness with poisoned whisky colored with the blood of its victims to the complexion of lurid brandy.[20]

Meanwhile, inside the theater proper:

> The entire inhabitants of houses of prostitution would customarily attend the theater in a body, entering the [third] tier by a separate stairway an hour or two before the rest of the house was opened. Unlike the higher class prostitutes who sat throughout the theater and met customers there by pre-arrangement through such means as newspaper advertisements, the lower class prostitutes of the third tier made the initial contacts with their customers in the theater itself.[21]

Music was an integral part of nineteenth-century popular theater, from the first minstrel shows through vaudeville and early musical plays, and songwriters understood that something quite different from the genteel Victorian song was needed for working-class, ethnically diverse audiences. Lyrics for the popular stage tended to be humorous, ironic, sometimes deliberately crude, never sentimental or didactic. Musically, songs were designed to be easily grasped at first hearing, with diatonic or pentatonic melodies often drawn from or patterned upon folk tunes. Songs of this sort could be learned by ear in a few hearings and passed on orally, with the aid of texts printed in songsters. As a result, they were not as profitable for publishers as were genteel songs written for the musically literate classes. As E. M. Wickes said in a how-to guide published in 1916:

> A stage song is about as valuable to a publisher as a straight comic song. The boys in the streets sing and whistle it, people like to hear it in the theatres, but the song does not sell. A stage song may be of a risqué type, or of such an ultra-cheap sentimental order that it fails to make any impression on the feminine heart.[22]

Yet the distinction between genteel parlor songs and rowdy stage songs was not absolute. Some parlor songs passed into oral tradition, were performed in the theater, and were printed in songsters, while some songs for the popular stage were brought out in piano-vocal scores and sung at home by the musically literate.

---

[18]For further information on this subject, see *Theatre for Working-Class Audiences in the United States, 1830–1880,* ed. Bruce A. McConachie and Daniel Friedman (Westport, Conn.: Greenwood Press, 1985), and Mary C. Henderson, *The City and the Theatre: New York Playhouses from Bowling Green to Times Square* (Clifton, N.J.: James T. White, 1973).

[19]McConachie, *Melodramatic Formations,* 144.

[20]As quoted in McConachie, *Melodramatic Formations,* 122.

[21]Claudia D. Johnson, "That Guilty Third Tier: Prostitution in Nineteenth-Century American Theaters," in *Victorian America,* 113.

[22]E. M. Wickes, *Writing the Popular Song* (Springfield, Mass.: Home Correspondence School, 1916), 34.

Throughout the nineteenth century, even before the minstrel show took formal shape, songs with protagonists impersonating members of a culturally marginalized national or ethnic group were mainstays of the popular stage. These songs usually depended on racial and ethnic stereotyping, often crude in character and sometimes with intent to ridicule. It must be remembered, though, that by midcentury, first-generation Americans and recent immigrants made up a substantial part of the audience for popular theater, particularly in New York and other cities,[23] and that this proportion increased in the following decades as the immigrant population swelled. Irving Howe sees the relationship between these people and the popular theater as a positive one:

> The popular arts came to serve as a sort of abrasive welcoming committee. . . . Shrewd at mocking incongruities of manner, seldom inclined to venom . . . they exploited the few, fixed traits that history or legend had assigned each culture. They arranged an initiation of hazing and caricature that assured the Swede, the German, the Irish, and then the Jew that to be noticed, even if through the cruel lens of parody, meant to be accepted.[24]

The popular stage and its songs could help create a sense of unity and common cause among such marginalized or disempowered groups, not only by allowing them to be depicted on stage but also by directing humor or satire against the more privileged classes, as when the protagonist of a comic song or skit would be an African-American or a recent European immigrant aping the dress, speech, and culture of America's elite classes.

Songwriters themselves understood the role of the popular stage in the social order of the day. Ned Harrigan, born in a poor Irish neighborhood of New York City, created a series of shows (with Tony Hart and David Braham) set in that city, with casts including Irishmen, Germans, Italians, and African-Americans; he once said that his songs "helped to lighten the toil of the working people and were, and are now, potential peacemakers in many a gathering where they calm the angry passions of the poor. . . . Make songs for the poor, and you plant roses among the weeds."[25]

Before the 1880s, the vast majority of immigrants to the United States had come from the British Isles, the German-speaking countries, and Scandinavia. Though each new group had to struggle for acceptance in the New World, it eventually became integrated into American life through its assimilation of the language, laws, and customs of the new country. Social equality came more slowly. But beginning in the 1880s, the arrival of millions of people from Central and Eastern Europe and the Mediterranean and the first migration of former slaves from the South to other parts of the country brought a different dynamic to America's social order, for these people differed more radically from the older population in physical appearance, language, religion, and cultural background than had earlier immigrants from northwestern Europe.

Henry James was taken aback by the vast numbers of these new and different Americans. Though insisting that he had always been comfortable with such people in their own countries during his years abroad, James felt a "chill" when he encountered them in the United States. "There is no claim to brotherhood with aliens in the first grossness of their alienism," he insisted. "The material of which they consist is being dressed and prepared, at this stage, for brotherhood, and the consummation, in respect to many of them, can not from the nature of the case be, in any lifetime of their own."[26] Soon organizations would be founded to keep these "aliens" in their proper place, and political careers would be built on promises to change the country's immigration laws.

---

[23]McConachie, at various points in *Melodramatic Formations,* discusses the constitution of theater audiences in the middle decades of the century.

[24]Irving Howe, *World of Our Fathers* (New York: Harcourt Brace Jovanovich, 1976), 402.

[25]E. J. Kahn, Jr., *The Merry Partners: The Age and Stage of Harrigan and Hart* (New York: Random House, 1955), 152–53.

[26]James, *The American Scene,* 120.

As problematic as this new multiculturalism may have been for Henry James and others, Tin Pan Alley thrived on it. Audiences for the popular musical theater swelled, fresh venues such as Tony Pastor's new Music Hall (1881) opened, and the popular theater began to shed its rowdy image. By the turn of the century, national chains of theaters, organized by booking agents, blanketed the country with popular entertainers from the larger cities, especially New York. The popular music industry—songwriters, performers, publishers, impresarios—came to be dominated increasingly by first-generation and immigrant Germans, Irishmen, Italians, and Jews, and the songs produced by this industry depended more heavily than ever on ethnic protagonists.

Nevertheless the distinction between genteel and stage songs persisted: sentimental, didactic ballads for home consumption were still written and published, and lively, humorous, often irreverent and sardonic show songs were still performed in vaudeville and minstrel shows and in America's nascent musical comedies. But in a process that would later be called "crossover," sales of sheet music arrangements of songs written for the stage increased as the growing respectability of the popular theater and the talent of its songwriters caught the attention of musically literate people from the "middling" classes. At the same time, these songs for the musical stage began drawing more on African-American styles and genres. Even after having been recast by Tin Pan Alley songwriters, the cakewalk, the so-called coon song, ragtime, and syncopated dance music still represented a new, brash, dynamic sensibility that Americans in increasing numbers and from different classes found refreshing and exciting.

The Tin Pan Alley era also brought intensified commercialization of popular music. Publishing houses, most of them in New York, developed new techniques of production, promotion, and marketing designed to maximize sales of sheet music. Soon it was not uncommon for a successful popular song to sell a million or more copies, with unprecedented profits accruing to the publisher and sometimes the songwriter. Coinciding as it did with the era of Henry Ford's assembly line and other revolutionary industrial production techniques, this commercial blossoming of the popular music industry sparked a reaction to so-called mass culture that was to persist throughout much of the twentieth century. As one critic put it in the 1950s, "Mass culture threatens not merely to cretinize our taste, but to brutalize our senses while paving the way to totalitarianism."[27] Resistance to mass culture brought together unlikely bedfellows: America's intellectual elite, much of the religious establishment, performers and composers of classical music, and eventually Marxist theorists and polemicists.[28]

Irving Berlin and Tin Pan Alley Song Production

Irving Berlin (1888–1989), born in Russia, emigrated with his family to New York City in 1893 and grew up in a Lower East Side Jewish neighborhood.[29] From a beginning as a singing waiter and a song plugger, he gradually worked his way up in the world of

[27]Bernard Rosenberg, "Mass Culture in America," in *Mass Culture: The Popular Arts in America* (New York: Free Press, 1957), 9. This anthology gives an excellent overview of the issues involved. For a more recent summary, see Charles Hamm, "Some Fugitive Thoughts on the Historiography of Music," in *Essays in Musicology: A Tribute to Alvin Johnson*, ed. Lewis Lockwood and Edward Roesner (n.p.: American Musicological Society, 1990), 284–91.

[28]See Macdonald Smith Moore, *Yankee Blues: Musical Culture and American Identity* (Bloomington: Indiana University Press, 1985), for a discussion of one group of writers and musicians, centered in New England, who argued that the moral and intellectual strength of the United States lay in its British-based cultural life, which they saw as threatened by new patterns of immigration and new popular cultural forms.

[29]The biographical literature on Berlin includes Alexander Woollcott, *The Story of Irving Berlin* (New York: G. P. Putnam's Sons, 1925); Michael Freedland, *Irving Berlin* (New York: Stein & Day, 1974) and *A Salute to Irving Berlin* (London: W. H. Allen, 1986); Ian Whitcomb, *Irving Berlin and Ragtime America* (London: Century Hutchinson, 1987); and Laurence Bergreen, *As Thousands Cheer: The Life of Irving Berlin* (New York: Viking, 1990).

Tin Pan Alley. Several songs written before he was twenty, including "No One Could Do It Like My Father!", "Sadie Salome (Go Home)," and "My Wife's Gone to the Country (Hurrah! Hurrah!)," enjoyed enough commercial success to earn him a position as a salaried songwriter at the Ted Snyder Company. National and then international fame came with a succession of hit songs culminating with "Alexander's Ragtime Band" in 1911, and in 1912 Berlin was made a partner in the publishing firm of Waterson, Berlin & Snyder.

His early songs were written in the context of Tin Pan Alley's cultural milieu. Most are suited for performance on the popular musical stage, many of them for specific shows or singers, others are ballads with sentimental lyrics set in retrospective musical styles. Thus the late nineteenth-century dichotomy between the repertory for the popular theater and that for the home circle persists in these songs.

They follow Tin Pan Alley practice in mode of production as well. In the first stage of creation, a songwriter, or often two or more of them working together, would toy with fragments of melody or text until something useful took shape. Theodore Dreiser described the work habits of his brother, Paul Dresser:

> He was constantly attempting to work [his songs] out of himself, not quickly but slowly, brooding as it were over the piano wherever he might find one and could have a little solitude, . . . improvising various sad or wistful strains, some of which he jotted down, others of which, having mastered, he strove to fit words to. He seemed to have a particular fondness for the twilight hour, and at this time might thrum over one strain or another until over some particular one, a new song usually, he would be in tears.[30]

Berlin proceeded in a similar way:

> I am working on songs all the time, at home and outside and in the office. I gather ideas, and then I usually work them out between eight o'clock at night and five in the morning. . . . I get an idea, either a title or a phrase or a melody, and hum it out to something definite.[31]

He jotted down fragments of lyrics as they came to him, on whatever scraps of paper were at hand. The most promising of these would be expanded and revised, and when one or more verses and/or a chorus had taken shape, a lyric sheet would be typed up and kept on file in his publisher's office. Like many other Tin Pan Alley songwriters, including Charles K. Harris, Berlin dictated the music of his songs to a staff pianist or arranger. First the melody would be written down as a lead sheet, without text, perhaps with a few chord changes indicated, then Berlin would work together with a staff musician on harmonization and accompaniment.[32] When a piano-vocal score had been completed and judged to be worthy of publication, it would be sent to an engraver. First-proof sheets from "punch" plates would be returned for correction,[33] and at this point prepublication "professional copies," without covers, might be circulated to potential performers. An artist would be commissioned to create a front cover, and the publishing house would design a back cover listing a number of its recent titles, giving a few bars from several new songs, or printing the entire first page of the chorus of one such song. Corrected proof of the music would be sent to the printer together with front and back covers, and copies would be run off for copyright deposit and distribution to retail outlets.

Berlin's songs were usually copyrighted after being printed in final form, not in a preliminary version or as a lead sheet.[34] Two copies would be sent to the copyright office

---

[30]Theodore Dreiser, "My Brother Paul," in *Twelve Men* (New York: Boni & Liveright, 1919), 96 and 99.

[31]Quoted in Bergreen, *As Thousands Cheer*, 57–58.

[32]LC-IBC contains a collection of handwritten lyrics. Some are early versions of published songs, some are complete but never published song texts, others are mere fragments. There are also lead sheets for a number of published and unpublished songs from 1912–13.

[33]This process is described in Wickes, *Writing the Popular Song*, 148–50.

[34]Exceptionally, two of Berlin's songs from this period, "It Can't Be Did!" and "Angelo," were copyrighted as printed lead sheets. Though Berlin seems not to have availed himself of the possibility, prepublication copyright could be secured. See Richard Crawford, "Notes on Jazz Standards by Black Authors and Com-

in Washington, D.C., where the piece would be assigned a copyright number in the E classification[35] and its title and other information entered in a handwritten register and on two 3″ × 5″ cards, one filed under the song's title and the other under the copyright claimant, usually the publishing house.

Protection within England and its colonies was obtained by sending two copies to the British Library in London for colonial copyright, and Canadian copyright would be obtained similarly, through Ottawa. Some of Berlin's early songs were engraved and printed in England as well. On a trip to London in October of 1910, Henry Waterson, who put up the initial capital for the Ted Snyder Company, signed a contract with Bert Feldman for the inclusion of selected songs from the Snyder catalogue in the highly successful Feldman's Sixpenny Editions. Feldman, who earlier had published songs by Charles K. Harris in his series, wanted more up-to-date American pieces, particularly ragtime songs. Beginning with "Yiddle, on Your Fiddle, Play Some Ragtime," Feldman brought out several dozen songs by Berlin, printed from newly engraved plates copied from the American editions.

Though the front cover for a song was usually completed before the two copies were sent to Washington, copyright was occasionally secured on a song with a provisional front cover giving only the title of the song or a list of pieces brought out by the publisher. In such a case, distribution and sale would be delayed until the artist delivered the front cover, but a second copyright wouldn't be secured on the song with its illustrated front cover. A second and even a third front cover would sometimes replace the original one at a later point in a song's publication history, and even more commonly a song selling well enough to warrant subsequent printings would be given a new back cover promoting pieces brought out by the publisher since the song's first printing. Replacement of front and back covers necessitated no new copyright, and thus it's unclear whether copyright protection on a song extended to its covers.

The most common type of front cover features a large, multicolored drawing relating to the song's content, with the song's title prominently displayed. Since cover artists were sometimes given incomplete or incorrect information about the song in question, the title on the cover often differs from that given at the head of the first page of the music. Thus Berlin's "I Want to Be in Dixie" is titled "I'm Going Back to Dixie" on the cover, and "Draggy Rag" becomes "Dat Draggy Rag." Throughout the present edition I've assumed that the title found in the music itself, not the cover, is the one intended by the songwriter.[36] Cover illustrations sometimes conflict with the sense of the song as well. Witness, for example, the white bandsmen on the cover of "Alexander's Ragtime Band" and the white pianist on the cover of "He's a Rag Picker."

Tin Pan Alley publishers were eager to link their songs to popular entertainers of the day. Most front covers include a small oval space reserved for the portrait of a performer who has sung or played the piece on the vaudeville stage or in a show. If another performer later took up the song, his or her portrait might replace that of the original artist in subsequent printings. In a general way, then, the commercial success of a song is reflected by the number of different performers whose portrait appears at one time or another on its front cover; in the case of a hit song such as "Alexander's Ragtime

---

posers, 1899–1942," in *New Perspectives On Music: Essays in Honor of Eileen Southern*, ed. Josephine R. B. Wright with Samuel A. Floyd, Jr. (Warren, Mich.: Harmonie Park Press, 1992), 245–90. As Wayne Shirley pointed out in a personal communication, prepublication copyright was most often obtained for pieces intended to be recorded but not necessarily published as sheet music, and since at this time Berlin expected to make money primarily from the sales of sheet music, it was convenient to wait until a song had been printed before sending it off for copyright.

[35]All songs submitted for copyright, whether in classical or popular style, were given a number in this class. The text of a song, without music, could be copyrighted only in the B (book) classification; this was almost never done.

[36]While direct communication took place between the songwriter and the engraver, including exchange of at least one set of proofs, the cover artist apparently sent his work directly to the engraver, and the songwriter may not have seen the cover of a given song until it was published.

Band," the number reaches several dozen. But if a song were intended from the beginning for one of the top stars of the day, a large portrait of that person might be the central feature of the original front cover.

After a song had been copyrighted and run off in quantity, no changes in text or music were made.[37] Any errors or imperfections remained uncorrected, no matter how popular the song became and how many subsequent printings were made. If Berlin felt that a song needed revision after publication, as happened with "Fiddle-Dee-Dee," "At the Devil's Ball," and "This Is the Life," new plates were engraved and another copyright obtained on the revised version. But aside from these several exceptions, at this stage of his career Berlin was more concerned with turning out new songs than in reworking older ones.

Style and Originality in Berlin's Early Songs

The history of Western classical music is often understood as a dynamic, linear progression of styles with a general movement toward greater complexity. Consequently, some musicologists and critics have tended to judge the historical importance of composers by the extent to which their music moves forward, breaking new stylistic ground. By the early twentieth century, the suggestion that a new classical piece sounded like the music of another composer could be meant (and taken) as negative criticism. Audiences were often resistant to new pieces written in more complex styles than they were accustomed to, but some composers and critics viewed audience rejection as an inevitable reaction to the stylistic progressiveness of the piece in question. Since differences among compositions can be described through harmonic, melodic, rhythmic, and structural analysis, such analyses have sometimes been used to underline the progressive features of certain "masterpieces" of the classical repertory, and their greater historical importance and artistic superiority over lesser pieces.

The perception of what separates a good Tin Pan Alley song from a poor one has been quite different from this. For songwriters such as Berlin, widespread approval by performers and audiences was the distinguishing mark of a good song, and a poor song was one no one wanted to hear again.[38] Since mass audiences were unlikely to respond favorably to a song that sounded too different from those they already knew, writing a good popular song required, first of all, the use of musical and textual materials already familiar to audiences. Popular songwriters weren't concerned with turning out products that moved beyond the style of their peers, but in working with them in a common idiom and establishing common ground with their audiences. Thus one cannot judge their songs according to whether or not they broke new harmonic, melodic, or structural ground, and audience rejection signaled failure, not success.[39]

All Tin Pan Alley songwriters understood this. More perplexing, now as then, is the question of why certain songs succeeded while others failed, even though they all appear to be remarkably similar to one another.

Since Berlin wrote more successful songs than any of his peers and was thus the "best" songwriter of his era according to the standards of the genre, one might be tempted to try to explain his superiority by identifying novel or unusually effective har-

---

[37]Copies of some of Berlin's songs sent to the British Museum for the purpose of securing copyright differ in some details from those found elsewhere. Lacking front covers and apparently printed up from uncorrected "punch" plates, they are the only examples of this stage of the production process that I've identified.

[38]This was not merely a matter of money, as Bergreen and other writers suggest. Almost everyone connected with Tin Pan Alley at this time was part of a community with a shared cultural identity, and writing a song that was accepted by members of this community was a contribution to it.

[39]As Alec Wilder writes about Berlin in *American Popular Song,* "Whatever idealism some of his songs have revealed, the core of his work has been eminently practical: his has been truly a body of *work*. . . . I think that to Berlin, as well as to many other song writers, a good song and a hit song are synonymous" (91–92).

monic progressions or innovative turns of melody and formal design. Such details can indeed be found in his music and will be commented on below, and equivalent twists abound in his lyrics. But one cannot be certain that these things account for the success of a given song, or of his career in general. At most, stylistic innovation served Berlin as seasoning, not as a basic ingredient, and his success as a songwriter is probably not explainable through conventional methods of stylistic analysis.

More promising is an approach that recognizes Berlin's unusual empathy with performers and audiences and seeks to explore how this empathy influenced his choice of textual and musical materials. The protagonists of most of his early songs are drawn from the community of working-class immigrants and first-generation Americans to which Berlin, his publishers and producers, his performers, and his audiences all belonged. The texts of these songs, written in the vernacular and sometimes accented English spoken by members of that community, offer scenarios drawn from the dramatic stuff of their everyday lives. Berlin knew all the music his audiences knew, and his songs make use of the common melodic, harmonic, and rhythmic patterns of this music and frequently offer direct quotations from one familiar piece or another. The same things could be said of other Tin Pan Alley songwriters, true enough. But Berlin, more effectively than any of his peers, drew on the collective knowledge and memory of his audience to fashion dramatic situations and musical phrases similar to those found in songs they already knew, shaped in slightly unexpected ways. His best songs were almost—but not quite—already known to his listeners when heard for the first time. They were old stories with new twists.

A Stylistic Overview of Berlin's Early Songs

The vast majority of Berlin's early songs consist of a brief piano introduction, several bars of vamp, two or more verses, and a chorus.[40] The introduction is most often drawn from the last bars of the chorus, though sometimes from the first phrase of the verse or chorus; the vamp usually anticipates the first several bars of the verse. As I wrote in an earlier study, "the skill and genius of Tin Pan Alley [songwriters] was revealed by what could be done within a tightly restricted formal structure, rather than by flights of fancy soaring to new and complex designs. One is reminded of similar restrictions embraced by writers of sonnets, by the Japanese poets of haiku verse, and by the great American bluesmen."[41]

Some of the earliest songs have multiple verses, in the style of the late nineteenth century: "No One Could Do It Like My Father!" has nine, for instance, and "My Wife's Gone to the Country (Hurrah! Hurrah!)" has eight. By 1910 two verses had become standard for published songs, though additional verses might be added in performance; some of these are found among unpublished lyrics in the Irving Berlin Collection and on period recordings.[42] Sometimes the verse is longer than the chorus, but usually the two are of equal length, most often sixteen bars in $\frac{4}{4}$ meter or thirty-two in $\frac{3}{4}$, $\frac{2}{4}$, or $\frac{6}{8}$. A handful of songs written through 1914, including "The Million Dollar Ball," "When the Midnight Choo-Choo Leaves for Alabam'," and "Down in

---

[40]These components of a song could be fitted together in various ways in performance. In one format, according to the evidence of period recordings, the introduction led to the vamp, played twice; the first verse was followed by the chorus, sung once or twice; after a return to either the introduction or the vamp, the second verse and another one or two renditions of the chorus would follow, and the performance would often end with a repeat of the introduction, as a coda. But other sequences were possible as well.

[41]Charles Hamm, *Yesterdays: Popular Song in America* (New York: Norton, 1979), 361.

[42]According to advertisements in trade journals, publishing houses would supply additional verses to performers on request. The Ted Snyder Company described Berlin's "After the Honeymoon" as "a comedy waltz song with 50 extra choruses" in *Variety*, 4 November 1911, and offered "any amount of extra verses" for his "Fiddle-Dee-Dee" in *Variety*, 26 July 1912.

Chattanooga," anticipate the dominance of chorus over verse that characterized Tin Pan Alley songs of the 1920s and after.

Verse and chorus tend to be of comparable musical interest as well as length, and in fact the most memorable melodic lines and lyrical inventions are often found in the verses of these songs. A performance of the chorus alone of any piece in this edition would compromise the intended effect of the song.

According to unequivocal directions in the sheet music, the chorus was to be sung twice after each verse and was thus heard four times in the performance of a two-verse song. This practice is verified by early cylinder recordings of Berlin's songs. On discs, however, with less time available, it's more common to find a single chorus sung after each of the two verses, or a single verse followed by two choruses. Even though the chorus usually conforms to the definition of "a refrain, or burden, of a strophic song, both text and music of which are repeated after each verse, or stanza, of changing text,"[43] some of these songs in fact have an entire second set of lyrics for the chorus following the second verse, and others have "catch lines" of alternate lyrics slotted into the final rendition of the chorus.

Both verse and chorus are often structured in four four-bar phrases, in such patterns as *ABAC* or *AA′BC*. From early on, though, Berlin tinkered with this design. "Yiddle, on Your Fiddle, Play Some Ragtime" contains a twenty-bar chorus with five rather than four phrases, in the pattern *ABACA′*, and other five-phrase choruses are found in "Angelo" and "Dog Gone That Chilly Man." The final phrase is extended by two bars to make an eighteen-bar chorus in many songs, including "Innocent Bessie Brown," "That Kazzatsky Dance," and "When You Kiss an Italian Girl." "Ephraham Played upon the Piano" has a chorus of only eight bars, and "That Mysterious Rag" is made up of three eight-bar phrases. Strikingly, these departures from foursquare patterns are found more often in songs with both words and music by Berlin than in pieces written with a collaborator. Asymmetric structures, found as early as "Dorando" of 1909, seem to have been part of Berlin's musical instinct.

All of Berlin's early songs are written in major keys. The vast majority are in C, F, G, B-flat, D, or E-flat, the choice made not for any apparent expressive or associative reason but to ensure that the notated melody fits on the treble clef, without ledger lines. Two early high-class ballads, "Just Like the Rose" and "Dreams, Just Dreams," were brought out in three different keys each, for low, medium, and high voice; all other songs in this edition were published in a single key, with the understanding that they could be transposed in performance.[44] Exceptionally, "Next to Your Mother, Who Do You Love?" was published first in the key of A-flat, then reissued from new plates in G, probably because the first key was considered awkward for amateur pianists.

Too brief for extensive modulation, many songs nevertheless move to other keys. Modulation to the dominant is frequent in internal phrases or at the end of the verse. Songs beginning in a minor key to suggest ethnic flavoring—"Marie from Sunny Italy," "Sadie Salome (Go Home)," and other novelty or "character" songs with Jewish or Italian protagonists—sooner or later shift to the relative or parallel major, where they end. Many ragtime songs, and a few others as well, have a chorus in the subdominant. And Berlin was fond of jumping to an unexpected key, often without modulation, in the second or third phrase of a verse or chorus. Typical in this regard are "One O'Clock in the Morning I Get Lonesome," with its unprepared drop to the flatted sixth at the beginning of the verse's third phrase, and "Keep On Walking," which shifts abruptly from C major to E major at the same spot.

---

[43] *The New Harvard Dictionary of Music*, ed. Don Randel (Cambridge: Harvard University Press, Belknap Press, 1986), 163.

[44] Performers did not have to make such transpositions themselves. Publishing houses advertised the availability of "professional copies" in various keys, in *Variety* and elsewhere.

The harmonic language of Berlin's early songs is tonal and triadic. Chromaticism occurs chiefly in the context of secondary dominants, with the V of V chord particularly favored. Often the approach to an internal or final cadence will move through a circle of two or three secondary dominant chords, a harmonic mannerism associated with the barbershop quartet style. Nonharmonic notes are most often passing tones, auxiliary notes and, most characteristically, anticipations. In song after song, a melodic phrase reaches its destination one half-beat before the left hand of the accompaniment moves to the next chord. When this anticipatory note is harmonized in the right hand of the piano by the appropriate chord for the ensuing strong beat, a momentary tonic/dominant clash or even a simultaneous cross-relation will result; both are found in "Stop That Rag (Keep On Playing, Honey)." And sometimes chords and entire passages falling outside any of these harmonic usages will appear, such as the strange and striking out-of-key opening of the verse of "That Mysterious Rag" and the sardonic, dissonant instrumental break in the second chorus of "Hiram's Band" that seems worthy of Charles Ives.

Melodies tend to move stepwise or to outline a triad, straying from the diatonic scale only for an occasional chromatic passing tone or a modulation. They generally remain within the range of an octave, a limitation stemming from the very nature of the genre. As Wickes explained:

> A popular tune to become really popular must be simple—easy to play, and easy to sing. . . . The majority of performers cannot do justice to a song that goes beyond an octave in range—a fact which the arranger knows from experience. This is the chief reason why composers aim to keep melodies within the octave, or within an octave and a note, for unless the rank and file of singers use the song it will not stand much chance of being heard by the public. . . . Long and frequent jumps [should also be avoided]. Whether you move forward or backward, try to build by easy stages.[45]

Berlin himself put it even more pointedly: "Since the song is to be sung by an average singer, the compass must not range beyond an octave. That is why some melodies sound so ordinary."[46]

Like other songwriters of the era, Berlin often quotes or paraphrases other music. Sometimes he draws on his own pieces, as in his many references to "Alexander's Ragtime Band," but more often he borrows from other composers. Though the stylistic range of quoted material is considerable—Stephen Foster's plantation songs, Italian opera, "high-class" ballads, hit songs by contemporary songwriters, classical piano works—the common denominator is that it all could be heard on the streets of New York City at the turn of the century.

Rather than being an intellectual exercise or a display of erudition, Berlin's quotation of other music is almost always a response to the lyrics, and its function is to establish a common bond with his audience. As the lyrics of "Abie Sings an Irish Song" tell how Abie begins singing "By Killarney's lakes and dells" whenever an Irish customer comes into his clothing store, the music quotes from Balfe's setting of "Killarney." Similarly, Berlin quotes from his own "Sweet Italian Love" when the text of "Pick, Pick, Pick, Pick on the Mandolin, Antonio" mentions "sweet Italian love." The piano introduction to "A True Born Soldier Man" quotes "America" and "The Girl I Left Behind" to establish that the protagonist is an American soldier, and the introduction to "My Sweet Italian Man [II]" begins with several bars from *Pagliacci* to set an appropriately somber mood for the song. As the lyric of "Stay Away from the Fellow Who Owns an Automobile" warns one to "say Goodbye forever" when the speed of a machine reaches sixty miles an hour, the music echoes the dramatic climax, on these words, of "Good-bye" (1881) by Paolo Tosti.

---

[45]Wickes, *Writing the Popular Song,* 100.

[46]Irving Berlin, "How to Write Ragtime Songs," *Ideas* (1912): 53, clipping in the first of Berlin's scrapbooks in LC-IBC.

In several songs Berlin's use of borrowed material is so extensive as to approach parody. "He Promised Me" toys with fragments of Reginald DeKoven's "Oh Promise Me" throughout; "Opera Burlesque" is nothing less than a fourteen-page reworking of the famous sextet from *Lucia di Lammermoor;* and the chorus of "Pickaninny Mose," not published until 1921 but sketched in his early years, echoes both text and music of Ethelbert Nevin's "Mighty Lak a Rose" in every phrase.

Berlin was not "stealing" the music of other composers in these and similar instances. As noted above, his early songs are set in contemporary New York City, and every aspect of their style was designed to make them easily accessible to their audiences. Just as the lyrics often mention famous people of the day such as Enrico Caruso, John D. Rockefeller, or Theodore Roosevelt, so the music quotes or paraphrases tunes likely to be known to his audiences, as part of Berlin's technique of packing a new song with material designed to make his listeners comfortable with it on first hearing.

Piano accompaniments range from simple oompah-pah or oompah patterns in waltz and march songs to much more complex and technically demanding passages elsewhere, including arpeggiated patterns and full chords in both hands. Despite such clumsy details as second-inversion chords or open fifths on strong beats, and doubled leading tones, these accompaniments serve their purposes well and many of them are more idiomatic and rewarding for the pianist than the simpler ones of the 1920s and 1930s.

Despite the surface stylistic homogeneity of these songs, they differ from one another in small and sometimes large ways. Some of the differences reflect style change between 1907 and 1914. The sixth degree of the scale gradually intrudes on the tonic chord, for instance, first as a nonharmonic tone and later as an added chord tone not requiring resolution. Melodic sequence, found first as a structural device in the opening phrases of verse or chorus, later begins to be used more to create forward movement and climax, as in the last twelve bars of "Always Treat Her Like a Baby." Accompaniments called less often for thick-textured chords, arpeggios, and other figurations associated with the solo piano literature, from early to late.

But changes in musical style over time can only go so far in accounting for differences among the songs. Even more important are differences conditioned by the several categories of Tin Pan Alley song, and much of the skill of songwriting lay in the ability to fashion pieces within the stylistic parameters of each type of piece. In order to understand more clearly the expressive range of Berlin's early songs and their meanings for audiences, to judge his skill as a songwriter, and to document his role in the history of Tin Pan Alley, these pieces must be discussed in the context of the categories or genres to which they belong.

Genre in Tin Pan Alley Song

Most writings on genre formation have been produced as tools for studying literature.[47] But *genre,* beyond any specific connotations, is a useful term for things sharing common traits, and it will be used here in this general sense.

My taxonomy of Berlin's early songs draws on earlier attempts at classification. Though publishers' advertisements for their songs, found in trade journals and on the

[47]To attempt even a partial bibliography of such studies is beyond the scope of the present essay. However, Alastair Fowler's *Kinds of Literature: An Introduction to the Theory of Genres and Modes* (Cambridge: Harvard University Press, 1982) offers a good general introduction to the topic as it was treated by literary critics up to the last quarter of the twentieth century and a basic bibliography; and John Snyder's *Prospects of Power: Tragedy, Satire, the Essay, and the Theory of Genre* (Lexington: University Press of Kentucky, 1991) contains useful if not always sympathetic summaries of more recent issues raised in the work of E. D. Hirsch, Jacques Derrida, Frederic Jameson, and Michel Foucault, among others. As a general strategy, I have heeded Snyder's admonition that "every work deviates from any particular set of characteristics that may be attributed to its kind, and over time every work combined with all others of more or less the same kind constitutes the history of the genre."

back covers of sheet music, offer a disorderly assortment of labels, E. M. Wickes in 1916 instructed the novice songwriter that all songs were either ballads or novelty songs, though with numerous subdivisions of each.[48] Wickes's twofold division corresponds in a general way to the two functional categories discussed above: ballads are for the home circle, novelty songs are for the popular theater. Isaac Goldberg, writing in 1930, elevated two of Wickes's subgroupings of novelty songs to the level of separate categories, proposing a fourfold classification: ballads, novelties, dance songs (rags and blues), and production numbers.[49] A recently published set of six folios of Berlin's songs is organized according to the songwriter's own wishes, as expressed in a letter to his lawyer in 1947. One folio is devoted to ballads, one to novelty songs, and another to ragtime and other early songs; there are folios for each of two categories of show songs (Broadway and film), and a sixth one bringing together a topical selection of patriotic songs.[50] My own fourfold grouping—ballads, novelty songs, ragtime and other dance songs, show songs—represents a consensus and consolidation of these earlier classifications and is consistent with each of them.

Classification is a more complex matter for songs than for poetry or prose, since lyrics, music, and performance context and style must all be considered. Though the starting point for the groupings offered here is the style and structure of music and lyrics, other factors have been considered as well: identity of the song's protagonist; the intended performance venue; and ideology, defined as the relationship of a song's expressive content to dominant social practice. Thus, while a ballad is defined chiefly by the style and structure of its music and text, a so-called coon song is distinguished first and foremost by the race of its protagonist, an urban novelty by its flouting of public morality, and a show song by its inclusion in a stage production. As a result of these flexible criteria for classification, many songs are placed in two or more different genres or subgenres. To take the most famous example, "Alexander's Ragtime Band" is classified as a ragtime song because of its title and certain musical features, as a coon song because of its black protagonist, and as a show song because of its performance in a specific stage production.

Within each genre or subgenre, songs are listed in the chronological order of their copyright.

I. Ballads

A ballad, as the term was understood by popular songwriters in the first decades of the twentieth century, was a lyric, nondramatic song projecting a romantic, pathetic, nostalgic, or moralizing sentiment. Of all Tin Pan Alley genres, it had the closest ties with the genteel tradition of the late Victorian era.

Texts elaborate on a single situation or emotional state, presenting a vignette rather than a dramatic narrative moving to a resolution. The Tin Pan Alley ballad thus differs from those poems and songs classified as ballads by literary critics on the basis of their folk or popular origin and their strophic, narrative texts, and also from the dramatic, narrative art songs of the nineteenth and early twentieth centuries purported to reflect folk culture. Moreover, the term *ballad* didn't necessarily imply a slow tempo, as was the case in jazz and later eras of popular song.

Isaac Goldberg wrote in 1930:

> Ballads, as a rubric, cover a multitude of sins. . . . Though the tempo of a ballad may vary from a slow waltz to a vigorous march, the type now favors a more leisurely pace; the words,

[48]Wickes, *Writing the Popular Song*, 6–35.
[49]Goldberg, *Tin Pan Alley*, 211–16.
[50]*The Songs of Irving Berlin* (New York: Irving Berlin Music Company, 1991).

rather than the music, determine the classification. Ballads center about the home (mother, dad, children), cabins, shacks, cottages for two . . . and later, more. They may be racial (especially Irish); they may be rustic.[51]

Faced with such a range of subject matter, style, and expression, one must resist further generalization about the ballad and discuss each of several subgenres in turn. Table 1 shows the wide range of Berlin's ballad writing.

1. HIGH-CLASS BALLADS

Publishers used the label *high-class ballad* for serious songs by such composers as Amy Beach, Ethelbert Nevin, Edward MacDowell, and Carrie Jacobs-Bond, and also for pieces by popular songwriters who, with little or no training in classical music, attempted to emulate the style.[52] Though it's been suggested that Berlin was inspired to write pieces of this sort by British music-hall songs heard during his first visit to England, in 1910,[53] he had every chance to hear high-class ballads in New York City.

Protagonists of Berlin's ballads of this sort are always the unnamed "I," addressing the likewise generic "you." Texts are self-consciously poetic in sentiment and vocabulary, often using words, phrases, and images lying outside everyday speech. The beginning of "Dreams, Just Dreams" is typical:

> The sunbeams have taken their flight,
> The day slowly turned into night;
> All nature's at rest,
> The birds in their nest,
> Sleep on 'neath the moon's silv'ry light.

Unlike most of his other songs, Berlin's high-class ballads have no vamp leading into the verse and no repeat of the chorus after each verse. A few are through-composed, dispensing with the ubiquitous verse-chorus structure. Melodies may extend a third or a fourth beyond the octave range observed in other genres and may move to dramatic climaxes on high notes. Harmonies often include augmented or diminished fifths and other types of chromaticism rarely found in other songs. Accompaniments may feature such pianistic devices as arpeggiated chords, tremolos, and octaves in both hands. Though the use of compound meters was thought by Wickes to "handicap the success of a song, as it places it beyond the singing ability of the ordinary vocalist,"[54] Berlin's "Dreams, Just Dreams" is written in $\frac{6}{8}$ and the verse of "That's How I Love You" is in $\frac{12}{8}$. Syncopated patterns suggesting popular dance rhythms are never present. According to Goldberg, high-class ballads were sometimes called "black and whites," since they were "usually printed between sober covers, as befits a labor of such dignity."[55] The cover of "Dreams, Just Dreams" has a black-and-white drawing with a classical sylvan motif; those of "My Melody Dream" and "God Gave You to Me" lack illustration of any sort.

2. ROMANTIC BALLADS

The lyrics of romantic ballads, written in vernacular rather than poetical language, deal with happy romantic relationships, though a few anticipate the later torch song with their complaints of unfulfilled love. Most of these texts are cast as first-person expressions of sentiment addressed to an unidentified "you," though in several instances—as

---

[51]Goldberg, *Tin Pan Alley*, 213.

[52]For more information on this genre, and particularly the role of British publishers in promoting public performances of ballads, see Wilma Reid Cipolla, "Marketing the American Song in Edwardian London," *American Music* 8 (Spring 1990): 84–94.

[53]See Whitcomb, *Irving Berlin and Ragtime America*, 72–74.

[54]Wickes, *Writing the Popular Song*, 10.

[55]Goldberg, *Tin Pan Alley*, 214.

Table 1. Ballads

Subgenre	Date	Item	Title
High-class	1909	9	Just Like the Rose
	1910	53	Dreams, Just Dreams
	1911	86	My Melody Dream
	1911	93	He Promised Me
	1912	103	Spring and Fall
	1912	108	That's How I Love You
	1912	121	When I'm Thinking of You
	1913	163	Take Me Back
	1914	180	God Gave You to Me
Romantic, duple time	1907	1	Marie from Sunny Italy
	1908	2	Queenie
	1908	3	The Best of Friends Must Part
	1909	17	Some Little Something about You
	1909	19	I Wish That You Was My Gal, Molly
	1909	20	Next to Your Mother, Who Do You Love?
	1909	25	Before I Go and Marry, I Will Have a Talk with You
	1909	27	Someone Just Like You
	1910	28	Telling Lies
	1910	57	Wishing
	1911	64	When I'm Alone I'm Lonesome
	1911	72	Molly, O! Oh, Molly!
	1911	75	You've Built a Fire Down in My Heart
	1911	81	There's a Girl in Havana
	1911	92	Bring Me a Ring in the Spring and I'll Know That You Love Me
	1912	129	Down in My Heart
	1913	148	You Picked a Bad Day Out to Say Good-bye
	1913	162	Kiss Your Sailor Boy Goodbye
	1913	A9	I Could Live on Love and Kisses
	1913	A11	I've Got a Lot of Love for You
	1913	A12	I've Got to Catch a Train, Goodbye
	1914	187	Furnishing a Home for Two
	1914	189	That's My Idea of Paradise
Romantic, triple time	1910	34	Dear Mayme, I Love You!
	1910	39	I Love You More Each Day
	1911	70	When It Rains, Sweetheart, When It Rains
	1911	78	After the Honeymoon
	1912	127	When I Lost You
	1912	132	If All the Girls I Knew Were Like You
	1913	150	We Have Much to Be Thankful For
	1914	178	If I Had You
March	1909	13	Good-bye, Girlie, and Remember Me
	1909	22	Christmas-time Seems Years and Years Away
	1910	47	Kiss Me My Honey, Kiss Me
	1911	82	Don't Take Your Beau to the Seashore
	1911	95	Yankee Love
	1913	152	Keep On Walking
	1913	157	Somebody's Coming to My House
	1913	170	Daddy, Come Home
	1914	171	This is the Life [I]
	1914	172	This is the Life [II]
Rustic	1911	61	Virginia Lou
	1913	149	Happy Little Country Girl
	1913	166	There's a Girl in Arizona
	1914	183	I Want to Go Back to Michigan (Down on the Farm)
	1914	190	When It's Night Time in Dixie Land
Domestic	1913	157	Somebody's Coming to My House
	1913	165	You've Got Your Mother's Big Blue Eyes!
	1914	185	Always Treat Her Like a Baby

Continued on next page

Table 1. *Continued*

Subgenre	Date	Item	Title
Rhythmic/vernacular	1910	49	Stop, Stop, Stop (Come Over and Love Me Some More)
	1911	83	Dog Gone That Chilly Man
	1911	87	You've Got Me Hypnotized
	1911	89	Bring Back My Lovin' Man
	1911	91	Cuddle Up
	1912	104	I've Got to Have Some Lovin' Now
	1912	123	Do It Again
	1912	137	Goody, Goody, Goody, Goody, Good
	1913	138	He's So Good to Me
	1913	145	Snookey Ookums
	1913	148	You Picked a Bad Day Out to Say Good-bye
	1913	167	If You Don't Want Me (Why Do You Hang Around)
	1913	A15	You're Goin' to Lose Your Baby Some Day
	1914	173	It Isn't What He Said, but the Way He Said It!
	1914	174	I Love to Quarrel with You

happens also in other genres—the verse is a third-person narrative introducing a protagonist who then speaks, in the first person, in the chorus. When identified, protagonists have Anglo-American names, with the exception of two Irish characters (both named Molly) and one Italian (Marie). Ethnic ballads were an important subgenre for some Tin Pan Alley songwriters, but Berlin wrote only these three. He must have agreed with Wickes's admonition that

> Irish ballads are a drug on the market, and the chances of making one popular are small indeed. . . . An Irish ballad must be exceptionally good to interest the mixed population of this country. While it is true that we have a large Irish element, it is also true that the majority of persons of native birth and foreign extraction of other countries are no more interested in Irish songs than they are in French or German songs.[56]

These comments are another clue that the ballad was associated chiefly with the culture of the older Anglo-Saxon mainstream.

Most romantic ballads appear to have been written for performance by amateurs at home, though some were sung on the vaudeville stage and a few were interpolated into shows. Musically, they differ from high-class ballads in that verses are usually introduced by a vamp, melodies move within an octave range and rarely climax on a high note, chromaticism is used sparingly, and accompaniments make fewer technical demands on the pianist.

Some two-thirds of Berlin's early romantic ballads are in duple time, usually with a time signature of **C** and a tempo marking of Moderato. None ranked high among his commercial successes or received much contemporary critical acclaim. Some lyrics approach parodies of sentimental ballads by other songwriters:

> I'd wish for a night in June,
> A silv'ry moon real soon;
> A moon that makes you want to spoon,
> And softly croon love's tune.
> ("Wishing")

Berlin's ballads in triple time are another matter. The romantic waltz ballad represented early Tin Pan Alley's most successful assimilation of the style and substance of

[56]Wickes, *Writing the Popular Song,* 16.

late Victorian song, yielding such classics as "Daisy Bell" (1892) by Harry Dacre, "My Wild Irish Rose" (1899) by Chauncey Olcott, "My Gal Sal" (1905) by Paul Dresser, and "Let Me Call You Sweetheart" (1910) by Beth Slater Whitson and Leo Friedman. Berlin showed an affinity for songs of this type almost from the beginning. Though not numerous, his early waltz ballads were more expressive and distinctive than those in duple time, and one of them, "When I Lost You," ranks among his greatest early commercial and critical successes and anticipates the flood of memorable waltz ballads to flow from his pen in later years.

3. MARCH BALLADS

Though their texts are similar to those of romantic ballads, march ballads project a different set of musical gestures. Written in $\frac{2}{4}$ rather than **C**, moving along briskly as demanded by a tempo marking of Marcia or Allegro, their accompaniments are punctuated by fanfares and other martial figuration. The notion of a romantic text sung in march style may appear curious today, and the subgenre had only a brief vogue, but Berlin's most successful ballad in duple time, as measured by sheet music sales and stage performances, was one of these, "Kiss Me My Honey, Kiss Me."

4. RUSTIC BALLADS

Sometimes called pastoral ballads, rustic ballads are distinguished from romantic ballads only by having texts concerned with nostalgia for childhood, first love, or a faraway home. Many of the classic songs of early Tin Pan Alley were of this sort, including "On the Banks of the Wabash" (1897) by Paul Dresser, "Where the Morning Glories Twine around the Door" (1905) by Harry Von Tilzer, "Down by the Old Mill Stream" (1910) by Tell Taylor, and "The Trail of the Lonesome Pine" (1913) by Ballard MacDonald and Harry Carroll. But the subgenre held little appeal for Berlin, either because of his preference for comic or irreverent sentiments or because his personal history prompted little nostalgia for his early years.

5. DOMESTIC BALLADS

The Tin Pan Alley domestic ballad, with its sentimental glorification of stable and happy family relations, echoes the didactic, genteel tone of Victorian song. Wickes subdivides this group further, into Mother Songs and Juvenile Songs.[57] Berlin practically ignored this subgenre, again probably because of his disposition and personal history.

6. RHYTHMIC/VERNACULAR BALLADS

Rhythmic/vernacular ballads deal with romantic relations in a way quite different from the romantic ballad. The music draws on characteristic rhythmic patterns of popular syncopated dances, and as period recordings verify, these songs were performed according to the tempo of such dance music, despite a ubiquitous tempo marking of Moderato. The tone of the lyrics is bantering or playful rather than sentimental, and the language is more colloquial:

> Bring back my lovin' man,
> Bring back my great big bunch of sweetness,
> Bring back them kisses sweet,
> Find 'em! Find 'em!
> The kisses with the steam behind 'em!
> ("Bring Back My Lovin' Man")

Berlin's rhythmic/vernacular ballads break important new ground for Tin Pan Alley. Previously, like other songwriters, he had introduced the rhythms of syncopated dance music into his songs only in connection with black protagonists or when the lyrics described the physical act of dancing. Here, though, the playful impudence and sensuality

[57] Ibid., 17–21.

of syncopated vernacular dance is linked to white protagonists, as are bantering texts implying a relationship between dance and sex and even terms of address ("honey" and "baby") previously serving as code words to identify black protagonists. A central dynamic of Tin Pan Alley song in the second and third decades of the twentieth century was the gradual assimilation into general usage of musical styles and modes of expression originating in black music or used by white songwriters to portray black protagonists, and Berlin's early rhythmic/vernacular ballads can be seen as important antecedents for what was to come later. One need only think ahead to such songs as "Ain't We Got Fun?" (Richard Whiting, 1921), "Yes Sir, That's My Baby" (Gus Kahn and Walter Donaldson, 1925), and any number of songs by George Gershwin and Berlin himself.

Berlin wrote both lyrics and music of all his songs in this subgenre, underlining again that the most distinctive songs of his early years were those written without a collaborator. It's not my intention to suggest that Berlin singlehandedly invented this kind of song. Pieces of a somewhat similar nature were written by other songwriters during the same period: "Honey Man" (1911) by Joe McCarthy and Al Piantadosi; "You Made Me Love You" (1913) by McCarthy and James V. Monaco; "I Wonder Where My Lovin' Man Has Gone" (1914) by Earl C. Jones, Richard Whiting, and Charles L. Cooke. But Berlin was unquestionably at the center of the early development of this subgenre.

II. Novelty Songs

The text of a novelty song sketches a vignette or a brief story of an amusing or provocative nature. Most commonly, the verse is a third-person narrative introducing the dramatic situation and the protagonist, then the chorus is sung in the first person by this protagonist, though some texts are in first person throughout. These pieces were written for performance on the musical stage, usually in vaudeville by singers noted for portraying characters of specific ethnicity or those finding themselves in certain comic or melodramatic situations, hence the alternate label of "character" song. As Isaac Goldberg sums it up, "Novelty songs . . . lend themselves to action, to mimicry, to histrionic effect. They are, unlike the ballads, songs that we listen to rather than sing ourselves, and usually the emphasis is comic."[58]

Novelty songs vary so much in text and music that the various subgenres must be considered separately. Their centrality and diversity among Berlin's early songs can be judged from table 2.

1. ETHNIC NOVELTY SONGS

Protagonists are drawn from one of the ethnic, national, or racial groups of New York City, always the implied setting of Berlin's songs of this period. Lyrics are written in dialect, usually in first person throughout, and protagonists behave in ways supposedly characteristic of their race or nationality. The result, particularly in Berlin's earliest ethnic novelties, is broad stereotyping in the tradition of the contemporary popular theater:

> My gal's pop he's nice wop,
> He's-a got-a much-a swell-a barber shop;
> Once he said when we wed,
> If the business is rotten he's a shave-a me for not'in'.
>
> ("Dat's-a My Gal")

The music is likewise sketched with broad strokes suggesting ethnic-specific stylistic features. German characters sing tunes resembling ländler, with beer-garden oompah-pah backing; Italians sing graceful melodies with accompaniments invoking plucked mandolins, tarantellas, or street serenades; Jews sing tunes in minor modes spiced with augmented seconds.

---

[58]Goldberg, *Tin Pan Alley*, 214.

TABLE 2. Novelty Songs

Subgenre	Date	Item	Title
Italian	1909	5	Dorando
	1910	29	Sweet Marie, Make-a Rag-a-time Dance wid Me
	1910	A2	Angelo
	1910	41	Sweet Italian Love
	1911	58	Dat's-a My Gal
	1911	73	When You Kiss an Italian Girl
	1912	97	Pick, Pick, Pick, Pick on the Mandolin, Antonio
	1912	107	Antonio
	1912	128	My Sweet Italian Man [I]
	1913	144	My Sweet Italian Man [II]
German	1909	31	Oh How That German Could Love
	1910	A1	It Can't Be Did!
	1910	50	Herman Let's Dance That Beautiful Waltz
	1910	55	Bring Back My Lena to Me
Jewish	1909	7	Sadie Salome (Go Home)
	1909	23	Yiddle, on Your Fiddle, Play Some Ragtime
	1910	45	Yiddisha Eyes
	1910	56	That Kazzatsky Dance
	1911	67	Business Is Business
	1911	85	Yiddisha Nightingale
	1912	112	Becky's Got a Job in a Musical Show
	1912	136	Yiddisha Professor
	1913	155	Abie Sings an Irish Song
	1913	160	Jake! Jake! the Yiddisher Ball-player
Coon	1909	12	Do Your Duty Doctor! (Oh, Oh, Oh, Oh, Doctor)
	1909	14	Wild Cherries
	1909	21	Stop That Rag (Keep On Playing, Honey)
	1909	24	I Just Came Back to Say Good Bye
	1909	26	That Mesmerizing Mendelssohn Tune
	1910	32	When You Play That Piano, Bill!
	1910	35	Grizzly Bear
	1910	37	That Opera Rag
	1910	40	Alexander and His Clarinet
	1910	42	Oh, That Beautiful Rag
	1910	48	Colored Romeo
	1911	59	That Dying Rag
	1911	60	Alexander's Ragtime Band
	1911	62	The Whistling Rag
	1911	74	Ephraham Played upon the Piano
	1911	76	Woodman, Woodman, Spare That Tree!
	1911	84	Ragtime Violin!
	1912	98	I Want to Be in Dixie
	1912	102	Opera Burlesque
	1912	114	When Johnson's Quartette Harmonize
	1912	119	Ragtime Soldier Man
	1912	126	When the Midnight Choo-Choo Leaves for Alabam'
	1912	134	Wait until Your Daddy Comes Home
	1912	A5	The Funny Little Melody
	1913	141	Anna 'Liza's Wedding Day
	1913	147	San Francisco Bound
	1913	154	The Pullman Porters on Parade
	1913	A16	That Humming Rag
	1914	186	He's a Rag Picker
	1914	190	When It's Night Time in Dixie Land
Rube	1912	115	Fiddle-Dee-Dee [I]
	1912	117	Fiddle-Dee-Dee [II]
	1912	125	Hiram's Band
	1913	149	Happy Little Country Girl
	1914	171	This Is the Life [I]
	1914	175	He's a Devil in His Own Home Town
	1914	176	This Is the Life [II]
Other ethnic	1912	101	Alexander's Bag-pipe Band
	1913	143	In My Harem
	1913	A10	I Want a Harem of My Own

Continued on next page

TABLE 2. *Continued*

Subgenre	Date	Item	Title
Urban	1909	4	I Didn't Go Home At All
	1909	6	No One Could Do It Like My Father!
	1909	8	My Wife's Gone to the Country (Hurrah! Hurrah!)
	1909	10	Oh, What I Know about You
	1909	11	Someone's Waiting for Me (We'll Wait, Wait, Wait)
	1909	15	Oh! Where Is My Wife To-night?
	1909	16	She Was a Dear Little Girl
	1909	18	If I Thought You Wouldn't Tell
	1910	36	Call Me Up Some Rainy Afternoon
	1910	38	I'm a Happy Married Man
	1910	43	Try It on Your Piano
	1910	44	"Thank You, Kind Sir!" Said She
	1910	46	Is There Anything Else I Can Do for You?
	1910	52	Innocent Bessie Brown
	1910	54	I'm Going on a Long Vacation
	1911	66	When You're in Town
	1911	71	Don't Put Out the Light
	1911	77	Run Home and Tell Your Mother
	1911	80	One O'Clock in the Morning I Get Lonesome
	1911	82	Don't Take Your Beau to the Seashore
	1911	94	Meet Me To-night
	1911	96	How Do You Do It, Mabel, on Twenty Dollars a Week?
	1912	99	Take a Little Tip from Father
	1912	109	I'm Afraid, Pretty Maid, I'm Afraid
	1912	112	Becky's Got a Job in a Musical Show
	1912	116	Call Again!
	1912	118	The Elevator Man Going Up, Going Up, Going Up, Going Up!
	1912	120	Keep Away from the Fellow Who Owns an Automobile
	1912	135	Don't Leave Your Wife Alone
	1913	142	Welcome Home
	1913	152	Keep On Walking
	1913	158	I Was Aviating Around
	1913	170	Daddy, Come Home
	1914	179	If You Don't Want My Peaches (You'd Better Stop Shaking My Tree)
	1914	184	If That's Your Idea of a Wonderful Time (Take Me Home)
Music	1910	32	When You Play That Piano, Bill!
	1910	51	Piano Man
	1911	74	Ephraham Played upon the Piano
	1912	102	Opera Burlesque
	1912	114	When Johnson's Quartette Harmonize
	1912	125	Hiram's Band
	1912	136	Yiddisha Professor
	1913	168	Tra-La, La, La!
	1914	186	He's a Rag Picker
Miscellaneous and topical	1911	63	That Monkey Tune
	1912	111	A True Born Soldier Man
	1912	122	Come Back to Me, My Melody
	1912	130	At the Devil's Ball [I]
	1912	133	At the Devil's Ball [II]
	1913	139	At the Devil's Ball [III]
	1913	146	The Apple Tree and the Bumble Bee
	1913	147	San Francisco Bound
	1913	151	The Ki-I-Youdleing Dog
	1913	153	The Old Maids Ball
	1913	154	The Pullman Porters on Parade
	1913	156	The Monkey Doodle Doo
	1913	169	Down in Chattanooga
	1913	A8	Down on Uncle Jerry's Farm
	1913	A13	Somewhere (but Where Is It?)
	1913	A14	The Tattooed Man
	1914	181	They're on Their Way to Mexico
	1914	182	The Haunted House
	1914	188	Stay Down Here Where You Belong

As Berlin matured as a songwriter, his ethnic characters became less stereotyped in speech, behavior, and musical style. For instance, both the melodramatic music and the passionately vengeful text of "Antonio" invoke turn-of-the-century Italian verismo opera more than the vaudeville stage:

> Antonio, don't you think that you can treat me so,
> Because I sharp-a da stiletto till she look-a much-a new,
> And pretty soon the people walk-a slow behind you.

And "My Sweet Italian Man [II]," with a piano introduction quoting several bars from *Pagliacci,* is more pathetic than comic: having left his wife behind in Italy when coming to America to earn money, the Italian protagonist dreams of a "nice kind fairy queen" who would turn him into a fish so he could swim back home to his wife and raise a family.

Songs with black protagonists, here labeled *coon songs* in keeping with terminology of the day, make up the largest and most complex subgenre of Berlin's ethnic novelties. Even though the popularity of such songs peaked around 1900, they continued to be written and performed through the first dozen or so years of the twentieth century, and songs with black protagonists, still sometimes performed in blackface, are found into the 1920s. Berlin's coon songs not only document the gradual transformation of this subgenre, they also reveal a dramatic change in his own treatment of black protagonists.

The identification of a protagonist as black is usually obvious from the lyrics and music of a song, though in some pieces it hinges on details in the text more telling then than now: proper names associated with black characters in minstrel shows and earlier coon songs (Alexander, Ephraham, Eliza); terms of address such as "honey" or "baby"; the use of "gwine" for "going"; the absence of the final *g* in "coming" and similar words; and other speech patterns thought to represent black usage. Front covers sometimes depict a coon song's protagonists as black, but just as often they do not. Period recordings may help, since singers assumed broad "negro" accents when portraying black characters, as in "That Mesmerizing Mendelssohn Tune" and "When the Midnight Choo-Choo Leaves for Alabam'."[59] But distinctions between white and black protagonists are increasingly blurred in texts of later songs, as are the musical styles used to portray black and white characters. In many of the later songs in the present edition, "Ragtime Soldier Man" and "San Francisco Bound" for instance, it's not clear whether the characters are black or white; and the music industry began to embrace this ambiguity by advertising that certain songs could be sung either as a coon song or not.[60]

Some of Berlin's earliest songs of this type portray black characters in ridiculous or demeaning situations, in the tradition of the genre. Liza Green visits her doctor in "Do Your Duty Doctor! (Oh, Oh, Oh, Oh, Doctor)" seeking to be cured of a "love attack." In "I Just Came Back to Say Good Bye," William returns to the home of his girlfriend, whom he's just abandoned, when he smells stew cooking in her kitchen. The musical instrument that excites Eliza in "Alexander and His Clarinet" is an obvious phallic symbol. But Berlin's later black characters are portrayed in more varied and sympathetic ways. The distinctive behavior of black protagonists in many of Berlin's later coon songs consists of making or enjoying music rather than engaging in buffoonery or theft, or wielding razors, and the music in question is not the ubiquitous banjo playing or fiddling associated with supposedly happy slaves and ex-slaves in "plantation" songs. Berlin's black protagonists play the piano or sing in quartets, and they are drawn as much to operatic music as to ragtime. The music itself may echo not only styles and genres commonly associated with African-Americans, such as ragtime and other

[59]As recorded in "negro" dialect by Arthur Collins and Byron G. Harlan, on Columbia A801 (Mx. 4328-2) and Indestructible (Columbia) Cylinder 3289, respectively.

[60]For instance, "When I'm Alone I'm Lonesome" was advertised in *Variety,* 4 November 1911, as "a wonderful ballad, or can be used as a 'coon' song."

syncopated popular dance styles, but also Italian opera, semiclassical piano music, and barbershop quartet singing.

Other later songs are concerned with a black mother coping in a single-parent home ("Wait until Your Daddy Comes Home"), a black wedding ("Anna 'Liza's Wedding Day"), and a black church service ("Revival Day"). And not only did Berlin portray black characters with more empathy than did most songwriters of the day, he even seemed to identify with two of them, Moses and Ephraham, who appear in a number of different songs.[61]

Though many of Berlin's ethnic novelties appear insensitive and even offensive today, they were written in a different social climate. Cumulatively, they carry the message, just beneath the surface of their stereotyping, jesting, and exaggeration, that American society—at least in New York City—was made up of peoples of varying ethnic and racial backgrounds, interacting with one another and finding some measure of common cause against those who would deny them full access to the American Dream.

By 1914 Berlin was writing fewer songs of this type, in response to a new social dynamic in the United States,[62] and the protagonists of many of these later songs were rural Anglo-Americans ("rubes") rather than more recent immigrants.[63]

2. URBAN NOVELTY SONGS

The protagonist of Berlin's "My Wife's Gone to the Country (Hurrah! Hurrah!)," a man whose wife and child have left the city for a vacation, celebrates his freedom by getting together with an old girl friend. As he puts it, "I love my wife, but oh! you kid, my wife's gone away." The song quickly sold some three hundred thousand copies of sheet music following its publication in June of 1909 and prompted immediate imitations by other songwriters ("I Love, I Love, I Love My Wife, But Oh You Kid" by Jimmy Lucas and Harry Von Tilzer, for instance) and an answer song later the same year by its own authors, "Oh! Where Is My Wife To-night?"

An urban novelty song develops a miniature drama in which some aspect of contemporary public morality is violated. Provocative flirtation, premarital sex, adultery, drinking, smoking, and even prostitution are detailed, or at least implied, in the lyrics.

Most commonly, a man seduces (or attempts to seduce) a woman, married or otherwise, often in a bar or café where dancing, smoking, and drinking are taking place. Often the woman turns the table in the second verse, by betraying the man or taking advantage of his attentions. In "She Was a Dear Little Girl," Betsy Brown, seduced by "the son of some millionaire," subsequently "introduces his check to his pen" to pay for her prodigious appetite and thirst when he takes her out to dinner. In another song, Bessie, just arrived from Kankakee, is initiated into the ways of the city by a man she meets "upon the avenue"; soon dozens of "fellows" are calling on her, making "a jew'lry store of Bess" ("Innocent Bessie Brown"). Elsewhere William C. Brown leaves his wife at home to go "aviating around":

> I was flying all over town,
> I struck a rathskeller, and broke my propeller,
> My feet wouldn't stay on the ground.
> I fell down, down, down where the Wurtzburger flows,
> And I thought I'd be drowned;
> I was saved by a queen, in another machine
> ("I Was Aviating Around")

---

[61]Charles Hamm, "The Early Songs of Irving Berlin As Biographical Documents," *The Musical Quarterly* 77 (Spring 1993): 19–22.

[62]Ibid., 9–15.

[63]In fact, over the next several years Berlin wrote more "rube" songs, including "I'm Going Back to the Farm" (1915), "Si's Been Drinking Cider" (1915) and "He's Getting Too Darn Big for a One-Horse Town" (1916), than any other type of ethnic novelty.

The next night his wife "wasn't there to receive him" when he came home; out "aviating" herself, she is "saved" herself, by a man she meets in a bar.

Sometimes titillation comes not from the actual commission of sins but from a warning to avoid a compromising situation, as in "Don't Put Out the Light" and "Keep Away from the Fellow Who Owns an Automobile" ("He'll take you far in his motor car, too darn far for your Pa and Ma").

None of this escaped self-appointed guardians of America's morality. Alexander Blume, writing in a New York evening newspaper, singled out Berlin's "My Wife's Gone to the Country" and "In My Harem" to support his contention that popular songs were contributing to the country's moral decay:

> If it is true, as has been said, that you may judge of a nation by its music, what a lot of shiftless, thoughtless and blatantly immoral people we must be, taking the "popular song" as our standard. . . . Did you ever hear a "popular song" that didn't laugh openly at the sacred institutions of marriage; that didn't frankly praise and encourage the faithlessness and deceit practiced by either friend, husband or wife? What has come over us anyhow? Decent women and girls with their men folks sit in theatres and applaud vociferously, amid their boisterous laughter, some singer who with well-studied indecency proceeds to gush forth songs of the most vulgar and immoral character. . . . There is, alas, sufficient tendency to ignore the responsibility that every decent man should feel incumbent upon him, without lending a hand to those who would shirk their bounden duties, by the proclamation of immoralities and subterfuges through these songs. . . . If I had my way I would appoint a rigorous censorship upon all so-called "popular songs," and make it a criminal offense to publish such songs as I have mentioned.[64]

Songs dealing with sex, drinking, gambling, and other forms of immorality had been written before this time, but always following one of two scenarios: either the protagonist was a member of some marginalized ethnic group, as in Hughie Cannon's "Bill Bailey, Won't You Please Come Home" (1902), a tale of marital strife and desertion in an African-American family; or the sin was condemned and punished, as in "A Bird in a Gilded Cage" (1901) by Arthur J. Lamb and Harry Von Tilzer, in which a young woman's greed in marrying a much older man for his money results in her disgrace and death. The protagonists in Berlin's urban novelties are drawn from the older mainstream of American society, though, and their moral transgressions are treated humorously. Thus these songs not only fail to offer models of socially acceptable behavior or condemn transgressions of this code, they also suggest that white, Protestant, even British-descended Americans—not only "aliens"—can and do violate the country's public standard of morality.

Wickes, after somewhat primly proposing the label of *suggestive songs* for these pieces, sets out to identify those responsible for their existence:

> The lewd lyric never was in favor with real song writers, for, apart from any moral question, they felt that they could get along without it. . . . Occasionally some actress with little regard for the public's morals, or even her own, has a suggestive song written to order. In the past when this took place there was always a publisher ready to bring the song out, depending upon the performer's reputation to bring him a few rot-rimmed dollars. Today, however, the average person has little time for performers of this character. The public has tired of having filth served up in the form of amusement.[65]

Berlin's songs of this sort form such a clearly defined group as to demand separation in my taxonomy, even though publishers didn't distinguish them from any other

[64]Quoted from a clipping from an unidentified newspaper article contained in a scrapbook kept by Berlin, now in LC-IBC. As might be expected, resistance to songs of this type was particularly strong in Boston. *Variety* for 27 June 1913 reported, for instance: "Following the Monday matinee at Keith's every act on the program with songs was informed that either one or more numbers used by them could not again be sung on that stage. The eliminated songs were all suggestive ones, and had proved during the afternoon show they were the best applause winners for each turn. . . . The action by the Keith management may be the commencement of a ban against suggestive lyrics, of which there have been a great number since ragtime songs became prevalent."

[65]Wickes, *Writing the Popular Song*, 27–28.

novelty songs. In this one instance the label (*urban novelty*) is my own, chosen because these songs are set in the city and because their content touches on cultural issues of the day revolving around urban life.

As Dennis Loranger sums up the growing conflict in values between country and city life in late nineteenth-century America, preindustrial family roles in which both sexes "were responsible for both the family's economic prosperity and emotional support" and the view of nature as "a harmonious helpmate to [their] efforts" still characterized rural life, at least as it was viewed from the city. In the new urban society, on the other hand, the male was usually the breadwinner while the female took on a greater nurturing role and also had more time for education, cultural activity, and other independent activities.[66] In the rustic ballads of some of Berlin's older contemporaries, Paul Dresser above all, rural family relations were portrayed as being in harmony with nature and were treated with wistful nostalgia, tragically lost to urban man, who was condemned to lead an unhappy, sinful, and unnatural life in the city. Berlin's urban novelties, on the other hand, celebrate the pleasures of city life, and when their protagonists are recent arrivals from farms or small towns, as is often the case, they show no desire to return to their rural roots.[67] Also, many of the female characters exhibit traits of the increasingly independent and assertive New Woman of the early twentieth century.[68]

One might expect the music of these songs to draw on styles associated with contemporary urban life, as do Berlin's rhythmic/vernacular ballads, but this is not usually the case. Syncopation is rarely present in either melody or accompaniment. Tunes are pervasively diatonic and sometimes even pentatonic, limited in range to an octave or less, laid out in foursquare phrases with the second phrase of the verse or chorus frequently a simple sequence to the first. Except for the ubiquitous secondary dominants and frequent passing modulations to the dominant, harmonies seldom stray from the basic triads of the tonic key. Without their texts, these pieces would strike the ear as somewhat old-fashioned. They are an odd but curiously effective mixture of a deliberately retrospective musical style wedded to aggressively modern lyrics, almost as though Berlin set out to create a body of contemporary urban folk songs.

The urban novelty was a product of the second generation of Tin Pan Alley songwriters. Charles K. Harris, Paul Dresser, and George M. Cohan wrote none of them, nor are precise models to be found in British music-hall songs of the day, which despite their humor and irreverence were less openly challenging to Victorian morality. "In My Merry Oldsmobile" (1905) by Vincent Bryan and Gus Edwards is one of the handful of pieces of this sort predating Berlin's first efforts, and among those contemporary with his own urban novelties are "Row, Row, Row" (1912) by William Jerome and Jimmie V. Monaco, "When I Get You Alone To-night" (1912) by Joe McCarthy, Joe Goodwin, and Fred Fisher, and "Let's All Go Around to Mary Ann's" (1913) by Ballard MacDonald and Harry Carroll.

3. MUSIC NOVELTY SONGS

Virtually alone among songwriters of his generation, Berlin took music itself as subject matter for novelty songs. He had a lifelong affection for the piano and thought of himself as a pianist, despite suggestions by some writers that he played poorly.[69] Most

---

[66]Dennis Loranger, "Women, Nature and Appearance: Themes in Popular Song Texts from the Turn of the Century," *The American Music Research Center Journal* 2 (1992): 70.

[67]As in Berlin's song "This Is the Life."

[68]See Loranger, "Women, Nature and Appearance," 68ff., and also Steven Mintz and Susan Kellogg, *Domestic Revolutions* (New York: Free Press, 1988).

[69]The rumor that Berlin played the piano poorly may trace back to a press conference in London in 1913, at which he allegedly played a new song with only one finger. It persisted until 1955, when he played the piano publicly in the course of a trial for plagarism. See Bergreen, *As Thousands Cheer*, 521–22.

songs in this group have piano-playing protagonists, usually black. The lyrics of many suggest that Berlin was passionately drawn to music and responded to it in an intensely emotional way.[70]

4. MISCELLANEOUS AND TOPICAL NOVELTY SONGS

Some pieces fit into no other subgenre of novelty song. "A True Born Soldier Man," for instance, is a "character" song with an ex-soldier who now "stays at home and fights with his wife" as protagonist. Several in this group (such as "They're on Their Way to Mexico" and "Stay Down Here Where You Belong") deal with topical issues, thus anticipating the political and patriotic songs that were to figure prominently later in Berlin's career. Still others refer to episodes in his own life: "The Haunted House" records his growing dissatisfaction with the publishing firm of Waterson, Berlin & Snyder, and "At the Devil's Ball" and "The Old Maids Ball" must have been inspired by specific social events.[71] Some songs in this group represent Berlin's contributions to one or another type of novelty piece popular at the moment, such as the "monkey" or "jungle" song. Given the heterogeneous nature of the songs grouped here, it's impossible to generalize about their stylistic traits beyond saying that Berlin's music is usually intended to fit the nature and substance of the text.

III. Ragtime and Other Dance Songs

Ragtime songs include all of Berlin's songs with the word *ragtime* in their titles or with mention of ragtime in their texts (see table 3).

Though "Alexander's Ragtime Band" was Berlin's most commercially successful song from these early years and the contemporary press endlessly referred to him as a writer of ragtime, his connection with what is known as *classic ragtime* is tenuous, at best.

Jasen and Tichenor give a good working definition of ragtime, as the genre has been understood since the ragtime revival began in 1950: "a musical composition for the piano comprising three or four sections containing sixteen measures each of which combines a syncopated melody accompanied by an even, steady duple rhythm."[72] They might have added that one of the internal sections, often labeled *trio,* is usually in the subdominant of the original key.

Three of Berlin's ragtime songs were originally piano rags by other composers, "Wild Cherries" and "Oh, That Beautiful Rag" by Ted Snyder and "Grizzly Bear" by George Botsford. Berlin transformed them into songs by adding a text and simplifying their structure: the first strain of the piano rag becomes the song's verse; the trio becomes the chorus, in the subdominant; other sections are discarded. Characteristic melodic and rhythmic patterns of piano ragtime, including tied and untied syncopations and secondary ragtime (three-note repeated or sequenced melodic patterns in the right hand clashing rhythmically with steady duple patterns in the left), are retained in Berlin's song versions.

In addition, three of Berlin's songs were converted into piano rags through a reversal of this process, whereby the melody of the verse becomes the first strain of the piano piece, the chorus becomes the trio, and an additional section is composed to bring the number of strains to three. "Alexander's Ragtime Band" and "The International Rag" were reworked into piano pieces by Berlin himself, apparently, and "That Mysterious

---

[70]See Hamm, "The Early Songs," for a fuller discussion of these issues.

[71]See ibid. for details.

[72]David A. Jasen and Trebor Jay Tichenor, *Rags and Ragtime: A Musical History* (New York: Dover, 1978), 1. The beginning of the ragtime revival is usually taken to coincide with the publication of Rudi Blesh and Harriet Janis, *They All Played Ragtime* (New York: Knopf, 1950).

Table 3. Ragtime and Other Dance Songs

Type	Date	Item	Title
Ragtime	1909	14	Wild Cherries
	1909	21	Stop That Rag (Keep On Playing, Honey)
	1909	23	Yiddle, on Your Fiddle, Play Some Ragtime
	1909	26	That Mesmerizing Mendelssohn Tune
	1910	29	Sweet Marie, Make-a Rag-a-time Dance wid Me
	1910	32	When You Play That Piano, Bill!
	1910	33	Draggy Rag
	1910	35	Grizzly Bear
	1910	37	That Opera Rag
	1910	42	Oh, That Beautiful Rag
	1911	59	That Dying Rag
	1911	60	Alexander's Ragtime Band
	1911	62	The Whistling Rag
	1911	79	That Mysterious Rag
	1911	84	Ragtime Violin!
	1912	88	Everybody's Doing It Now
	1912	100	Ragtime Mocking Bird
	1912	101	Alexander's Bag-pipe Band
	1912	113	The Ragtime Jockey Man
	1912	119	Ragtime Soldier Man
	1913	159	They've Got Me Doin' It Now
	1913	161	The International Rag
	1913	164	They've Got Me Doin' It Now Medley
	1913	A16	That Humming Rag
Other Dance	1910	50	Herman Let's Dance That Beautiful Waltz
	1910	56	That Kazzatsky Dance
	1912	105	Society Bear
	1912	106	Lead Me to That Beautiful Band
	1912	124	A Little Bit of Everything
	1913	156	The Monkey Doodle Doo

Rag" by William Schulz. The chorus of each of these songs is in the subdominant, but none of the three, including "Alexander's Ragtime Band," uses the rhythmic patterns of classic piano ragtime.

Beyond these six pieces, the connection between Berlin's ragtime songs and the style and structure of piano ragtime ranges from fleeting to nonexistent. Several of them written in collaboration with other composers ("Stop That Rag" and "That Dying Rag") exhibit hints of the rhythmic patterns of piano ragtime; the choruses of some are in the subdominant, or begin with a phrase in that key; and the rest do little more than mention ragtime in their titles or texts. "That Mesmerizing Mendelssohn Tune," for instance, has no discernible relationship to ragtime beyond the subtitle "Mendelssohn Rag" found on the sheet music cover (but not on the first page of the music). And stylistic issues aside, the mere number of songs by Berlin classifiable in any way as ragtime totals no more than two dozen, a small fraction of his output.

How, then, did Berlin's name come to be associated so closely with ragtime? The spectacular commercial success of "Alexander's Ragtime Band" was a factor, of course, as was his publisher's strategy of linking Berlin's name with the genre. More importantly, though, today's perception of ragtime stands at odds with history. As Edward A. Berlin argues:

> Had *Billboard* made a survey of favorite rags in 1902, the list probably would have included . . . all songs! Today, in contrast, ragtime is generally thought of as piano music. In a sampling of 230 ragtime-related articles and books from [1896 to 1920], only 21 refer to piano music, with a mere 16 citing specific piano rags. Since piano ragtime accounted for only a small part,

perhaps less than 10 percent, of what the music's contemporaries understood by the term "ragtime," it is necessary to consider the other forms as well.[73]

He elaborates on this argument elsewhere:

> It seems that the great ragtime successes were not piano pieces at all, but songs—such as *Alexander's Ragtime Band, Waiting for the Robert E. Lee, Hitchy-Koo,* and *Hello! Ma Baby.* When Rupert Hughes, Hiram K. Moderwell, and countless others wrote about ragtime, their concern was with songs; when pianist Mike Bernard performed in the Tammany Hall ragtime contest, he won it playing songs; when composer Charles Ives directed his keen and unbiased ears toward the popular music temper (captured and transformed in such works as his Study no. 18 for piano), he heard songs.[74]

One must remember also that ragtime was defined as much or more by performance style as by compositional details. Wickes wrote in 1916 that "now, everything that carries the jerky meter, or an irregular meter that possesses a pleasing lilt, is called ragtime."[75] Ragtime manuals demonstrate how popular and classical pieces could be played in ragtime style,[76] and Wickes observes that "a clever pianist can 'rag' the most sacred song ever published."[77] Many of Berlin's songs not included in this subgenre because their titles and texts make no reference to ragtime, such as "When the Midnight Choo-Choo Leaves for Alabam'," were sometimes perceived as ragtime songs because of the style in which they were sung and played. British journalists identified "I Want to Be in Dixie" as "one of his most popular rag songs" when he performed it in London,[78] and Berlin himself quotes both "When the Midnight Choo-Choo" and "I Want to Be in Dixie" as ragtime pieces in "They've Got Me Doin' It Now Medley."

To make sense of the contemporary perception of Berlin as a ragtime writer, then, one must consider not only the songs in this subgenre but also his other pieces marked by syncopation—his rhythmic/vernacular ballads and many of his novelty songs—and even certain romantic ballads performed in "jerky" style.

Wickes recalls that "for some time after the introduction of ragtime, only songs having to do with the negro were looked upon as being ragtime numbers."[79] Most of Berlin's early ragtime songs do have black protagonists and thus overlap with his coon songs, though exceptions can be found: "Yiddle, on Your Fiddle, Play Some Ragtime," "Sweet Marie, Make-a Rag-a-time Dance wid Me" and "Draggy Rag." With "That Mysterious Rag," copyrighted in the late summer of 1911, Berlin reached a critical turning point: neither the text nor the music of this or any of his later ragtime songs clearly implies a black protagonist. And at just about this time, Berlin began preferring the word *syncopated* over *ragtime* in connection with most of his own pieces. His assimilation of elements once associated with African-American music was well underway.

In summary, even though Berlin reshaped several piano rags by other composers into songs, and even though he was skilled at both textual and musical syncopation from early in his career, none of his own music bears more than a passing resemblance to the musical style of *classic ragtime.*[80] Nevertheless, his contemporaries considered him a master of ragtime.

---

[73]*Ragtime: A Musical and Cultural History* (Berkeley and Los Angeles: University of California Press, 1980), 2.

[74]Edward A. Berlin, "Ragtime Songs," in *Ragtime: Its History, Composers, and Music,* ed. John Edward Hasse (New York: Schirmer, 1985), 70.

[75]Wickes, *Writing the Popular Song,* 33.

[76]Ben Harney, *Ben Harney's Rag Time Instructor* (Chicago: Sol Bloom, 1897).

[77]Wickes, *Writing the Popular Song,* 33.

[78]*The Encore* (London), 3 July 1913.

[79]Wickes, *Writing the Popular Song,* 33.

[80]Edward Berlin explains that this term "had its origins in the ragtime period and was used by publisher John Stark and composer Scott Joplin to denote a music of superior artistic quality, a ragtime worthy of serious consideration." ("Ragtime Songs," 186). As Richard Crawford pointed out in a personal communication, the term *classic ragtime* seems always to refer to piano music, never to songs, both in the early twentieth century and in more recent literature.

Meanwhile, Berlin wrote a handful of pieces making reference in their titles or texts to dances other than ragtime. Table 3 distinguishes between these dance songs and Berlin's ragtime songs. As with ragtime songs, the texts are often invitations to the dance in question, and the music may suggest its characteristic rhythms and melodies.

IV. Show Songs

Show songs comprise all early songs by Berlin known to have been performed in a specific stage show or production. Since the unifying factor is function, this is the most stylistically heterogeneous group in my taxonomy, and in fact virtually all of these pieces have already been assigned a place in another genre. Table 4 includes all of Berlin's songs in this edition known to have been performed in specific shows.

For most of the nineteenth century, songs sung on the popular stage by professional entertainers functioned as discrete pieces in a string of events with little or no dramatic coherence. Even though the minstrel show followed an increasingly stylized structure as the century wore on, individual segments, including songs, remained interchangeable, with no relationship to each other or to any overall dramatic or musical scheme.[81] In the later nineteenth century and the first decades of the twentieth, vaudeville offered a succession of unrelated acts by different entertainers; any song of any sort could be sung at any point in a show, so long as its lyrics and music were appropriate for performance for a theater audience. Put a bit differently, minstrel and vaudeville songs weren't tailored for a specific show, nor did they bear any relationship to other songs in a show; they were interchangeable parts in a mosaic of miscellaneous episodes of singing, dancing, and comedy.

In many cases it isn't clear whether Berlin wrote a given song specifically for the show in which it appeared, or simply took it from his supply of unpublished or unfinished songs. Also, many songs not found in this group were sung on stage in day-to-day vaudeville, as opposed to featured, titled productions, and probably others were written with hopes of stage performance. Not surprisingly, then, no common stylistic features characterize these songs beyond those generally distinguishing stage songs from ballads. More extroverted than ballads for the home circle, they were projected to an audience by a singer backed by a pit orchestra. Their melodies sometimes extend beyond the usual octave range, to take advantage of the vocal talents of a particular singer. Didactic or maudlin sentiments are avoided in favor of humor, irony, satire, and irreverence. Lyrics are either narrative or addressed to the collective "you" (the theater audience), as opposed to the one-to-one expression of most ballads, and show songs are more likely to have extra choruses or additional "catch" lines to be inserted in later choruses. Front covers usually identify the show in which the song was performed and are often dominated by a large portrait of the featured singer.

Some of the songs in this group were performed as *production numbers*, a term best understood in historical context. Beginning in the 1870s with the extended theatrical sketches of Edward Harrigan, Tony Hart, and David Braham, continuing with turn-of-the-century operettas by Gustave Kerker, John Philip Sousa, and Victor Herbert and the musical plays of Will Marion Cook and George M. Cohan, and reaching fruition with the Princess Theatre shows written by Jerome Kern, Guy Bolton, and P. G. Wodehouse, a new theatrical genre took shape, musical comedy, in which spoken dialogue, song, dance, costume, and stage design are integrated into a single, continuous production.[82] The same stage characters are present throughout an entire show, their antics are governed by at least the semblance of a plot, and even though spoken dialogue,

---

[81]See Robert C. Toll, *Blacking Up: The Minstrel Show in Nineteenth-Century America* (New York: Oxford University Press, 1974) for the best general history of the minstrel show.

[82]For more information on the history of the musical stage and a bibliography on the subject, see Charles Hamm, "Musical Theater," *The New Grove Dictionary of American Music*, ed. H. Wiley Hitchcock and Stanley Sadie (London: Macmillan, 1986).

Table 4. Show Songs

Date	Item	Title	Show
1909	16	She Was a Dear Little Girl	*The Boys and Betty*
1909	21	Stop That Rag (Keep On Playing, Honey)	*The Jolly Bachelors*
1910	29	Sweet Marie, Make-a Rag-a-time Dance wid Me	*The Jolly Bachelors*
1910	30	If the Managers Only Thought the Same As Mother	*The Jolly Bachelors*
1910	31	Oh How That German Could Love	*The Girl and the Wizard*
1910	A1	It Can't Be Did!	*Jumping Jupiter*
1910	A2	Angelo	*Jumping Jupiter*
1910	35	Grizzly Bear	*Ziegfeld Follies of 1910*
1910	37	That Opera Rag	*Getting a Polish*
1910	41	Sweet Italian Love	*Up and Down Broadway*
1910	42	Oh, That Beautiful Rag	*The Jolly Bachelors*, *The Girl in the Kimono*, and *Up and Down Broadway*
1910	44	"Thank You, Kind Sir!" Said She	*Jumping Jupiter*
1910	47	Kiss Me My Honey, Kiss Me	*Up and Down Broadway* and *Jumping Jupiter*
1910	50	Herman Let's Dance That Beautiful Waltz	*The Girl and the Drummer* and *Two Men and a Girl*
1910	54	I'm Going on a Long Vacation	*Are You a Mason?*
1910	55	Bring Back My Lena to Me	*He Came from Milwaukee*
1910	57	Wishing	*The Girl and the Drummer* and *Two Men and a Girl*
1911	60	Alexander's Ragtime Band	*Friar's Frolic of 1911*
1911	65	I Beg Your Pardon, Dear Old Broadway	*Gabby*
1911	66	When You're in Town	*A Real Girl*
1911	68	Spanish Love	*Gabby*
1911	69	Down to the Folies Bergere	*Gabby*
1911	74	Ephraham Played upon the Piano	*Ziegfeld Follies of 1911*
1911	75	You've Built a Fire Down in My Heart	*Ziegfeld Follies of 1911* and *The Fascinating Widow*
1911	76	Woodman, Woodman, Spare That Tree!	*Ziegfeld Follies of 1911*
1911	79	That Mysterious Rag	*A Real Girl*
1911	80	One O'Clock in the Morning I Get Lonesome	*A Real Girl*
1911	81	There's a Girl in Havana	*The Never Homes*
1911	82	Don't Take Your Beau to the Seashore	*The Fascinating Widow*
1911	83	Dog Gone That Chilly Man	*Ziegfeld Follies of 1911*
1911	90	Sombrero Land	*La Belle Paree*
1911	91	Cuddle Up	*A Real Girl*
1912	98	I Want to Be in Dixie	*She Knows Better Now*, *The Whirl of Society*, and *Hullo, Ragtime!*
1912	100	Ragtime Mocking Bird	*She Knows Better Now*
1912	101	Alexander's Bag-pipe Band	*Hokey-Pokey*
1912	102	Opera Burlesque	*The Whirl of Society* and *Hanky-Panky*
1912	105	Society Bear	*The Whirl of Society*
1912	106	Lead Me to That Beautiful Band	*Cohan and Harris Minstrels*
1912	110	The Million Dollar Ball	*Hanky-Panky*
1912	113	The Ragtime Jockey Man	*The Passing Show of 1912*
1912	119	Ragtime Soldier Man	*Hullo, Ragtime!*
1912	124	A Little Bit of Everything	*Ziegfeld Follies of 1912*
1912	125	Hiram's Band	*The Sun Dodgers*
1912	129	Down in My Heart	*The Little Millionaire*
1912	131	Follow Me Around	*My Best Girl*
1913	140	At the Picture Show	*The Sun Dodgers*
1913	155	Abie Sings an Irish Song	*All Aboard*
1913	156	The Monkey Doodle Doo	*All Aboard*
1913	157	Somebody's Coming to My House	*All Aboard*
1913	163	Take Me Back	*All Aboard*
1913	167	If You Don't Want Me (Why Do You Hang Around)	*The Trained Nurses*
1914	172	Follow the Crowd	*Queen of the Movies*
1914	177	Along Came Ruth	*Along Came Ruth*
1914	187	Furnishing a Home for Two	*The Society Buds*
1914	189	That's My Idea of Paradise	*The Society Buds*
1914	190	When It's Night Time in Dixie Land	*Watch Your Step*

singing, and dancing usually occur separately, there are also sequences in which all three are combined into an integrated spectacle that came to be known as a production number.

Berlin didn't write a large collection of songs for a single stage show until late 1914, with the revue *Watch Your Step*, and not until 1925 did his first musical comedy, *The Cocoanuts*, reach the stage. Nevertheless, the integration of song, dance, and stage spectacle into production numbers was becoming a feature of variety shows, revues, and follies as well as musical comedies during the first two decades of the century, and Berlin wrote pieces of this sort for such shows as the early Ziegfeld Follies and the extravaganzas of Weber and Fields.

We know Berlin's show songs only from their published versions as vocal solos with piano accompaniment. No orchestral scores or parts have been preserved,[83] nor any indications for their staging; so it's often difficult to determine which songs were in fact staged as production numbers. "Opera Burlesque" and "Hiram's Band" both retain chorus parts in their sheet music versions, and sometimes a playbill, a photograph, or a review of a show will indicate that a certain song was done as a production number. We know, for instance, that "The Monkey Doodle Doo" was performed in Weber and Fields' *All Aboard* by an ensemble including chorus girls, each with a live monkey perched on her shoulders. But often production numbers cannot be distinguished from other show songs.

None of Berlin's early production numbers was among his best-known or best-selling songs. As Isaac Goldberg explains the problem:

> [Production numbers] are, in a double sense, show-pieces. They are intended for stage production, and while they may win great applause they do not attract purchasers in vast multitudes. Generally they are, as music, of a higher standard than the typical Broadway tune. . . . The critics praise your work, but the buyer keeps away.[84]

Performance Practice and Period Recordings

The present edition, like the first-edition scores on which it's based, offers these songs as vocal solos with piano accompaniment. As in the originals, pitches and rhythms are precisely notated, dynamic and tempo markings are given, and articulation is sometimes suggested by slurs, staccato dots, accent marks, and even occasional pedal marks in the piano accompaniment.

Performances of these songs today may take one of three general approaches.

The classical repertory is usually performed in a literal reading from the notated score, with variation between one performance and another limited to shadings and nuances of tempo, dynamics, timbre, and articulation. Berlin's early songs may also be performed in this manner today; such a reading can give some sense of their style and substance, and it has historical precedence in that the songs were done this way in the early twentieth-century home circle, though never in public.

A second option is to attempt to recover and follow period performance conventions. Some documentary evidence is available to bridge the gap between the notated versions of the songs and their public performance. We know, for instance, that they were sung to the accompaniment of a theater orchestra rather than a piano, on stage and in the recording studio.[85] Also, they were often performed by several singers rather than a

---

[83]Exceptionally, an orchestration of "Opera Burlesque" attributed to William Schulz is preserved in LC-IBC.

[84]Goldberg, *Tin Pan Alley*, 212.

[85]Publishers kept sets of instrumental parts on hand for sale or rental. These were often copied by hand within the publishing house, rather than being printed, and not many of these seem to have survived. The Harris Collection in Brown University's John Hay Library has a set of printed parts for an orchestral accompaniment for "Alexander's Ragtime Band."

soloist, most often as duos in which the two voices sometimes sang alternate phrases, sometimes sang together in harmony, or sometimes had one voice singing patter "fills" against a sustained note or rest in the other.[86]

Period recordings are the most useful evidence of contemporary performance practice. Though no comprehensive discography exists for this period,[87] at least a third of the songs in this edition are known to have been recorded around the time they were written. The following observations are based on fifty-odd period recordings made available to me through the generosity of Paul Charosh and Eric Bernhoft.

Concerning tempo, the rubric *moderato*, found in virtually every song in $\frac{4}{4}$ meter, often results in a pace of around M.M. 128, though the range extends from M.M. 100 to M.M. 150. The occasional song in $\frac{2}{4}$, usually a march ballad, tends to move at a slightly faster speed, with a more heavily accented downbeat. The few songs in cut time (¢) are performed with a pulse on the half note moving at about M.M. 190. No generalizations can be made for songs in $\frac{3}{4}$, which are all waltzes of one sort or another: "Herman Let's Dance That Beautiful Waltz," with its German protagonist, is a ländler (Tempo di Valse), "Dear Mayme, I Love You!" (Valse lento espressivo) and "When I Lost You" (Valse moderato) are love songs, moving at a slower pace as appropriate for the sentiments of their texts.

Dynamic markings appear sporadically and inconsistently. Some songs have only a single rubric at the beginning, some have one for the verse and another for the chorus, others are sprinkled more liberally with indications of changing dynamic levels, including such refinements as crescendo markings. In actual performance, though, dynamic level was a function of venue. Professional singers on vaudeville stages or in other large theaters needed to be heard throughout the house; since they sang without amplification and against an orchestra of a dozen or more players, low dynamic levels were not an option. Period recordings reinforce the notion that dynamic level was more a practical than an artistic matter. Almost without exception, the singer proceeds at a relentless forte or fortissimo, simply because nothing else would have recorded well on the equipment of the day. The accompaniment never overwhelms the voice, or even competes for attention; the orchestra plays at a lower dynamic level or is simply positioned further away from the recording horn.

The practical application of these observations to performance today would seem to be that dynamic markings in the score, like tempo indications, should be taken as no more than suggestions, to be overridden by the necessity of pitching the voice at a dynamic level loud enough to be heard clearly throughout the performance space, and of having the accompaniment subordinated to the voice.

As noted above, the protagonists of these songs may be ethnic minorities (Italian, Jewish, German, Irish, African-American, rural Anglo-American), or white, middle-class urban Americans, or unnamed first-person confessors of romantic sentiments, or the singer herself addressing the audience directly. The songwriter established this identity in the song's text by the choice of proper names and the use of dialect and stereotypical patterns of speech and behavior; it may also be underscored in the music by melodic, harmonic, and rhythmic patterns thought to be appropriate for such a protagonist. Far and away the most valuable lesson to be learned from period recordings is the extent to which singers shaped their delivery of text and music to underline, clarify, and project this identity.

In depicting a black, Jewish, Italian, or German protagonist, for instance, singers not only make use of whatever dialect is already written into the text, but also draw on ste-

[86]Among the materials in LC-IBC are holograph (or typed) lyric sheets, lead sheets, or both, of "double" versions of many songs; these are mentioned and sometimes quoted in the critical notes of the present edition.

[87]Despite its title, Brian Rust's *The Complete Entertainment Discography from 1897 to 1942* (New Rochelle: Arlington House, 1973) does not contain a comprehensive listing of recordings from the period.

reotypical ethnic pronunciation and speech patterns for the entire lyric; they may also interpolate exclamations, phrases and even entire sections of dialogue supposedly characteristic of the protagonist's ethnicity. At the other end of the class spectrum, the singer of a high-class ballad will enunciate each word carefully in "cultured" pronunciation, complete with rolled r's and touches of a British accent.

Ethnic and class depictions govern vocal style as well. High-class ballads are sung precisely on pitch, with a vocal production invoking the concert or recital hall rather than the vaudeville stage; legato phrasing predominates, whether the printed score calls for it or not, and high notes are sometimes sustained for their sheer sound. Novelty songs, on the other hand, are often more recited than sung, with little attention paid to precise pitch; the appropriateness of the label *coon shouter* is evident in recordings by such performers as May Irwin.

Performers went to such lengths to project the identity of their protagonists simply because this identity was often crucial to the meaning of a song. For instance, in a time-honored tradition of the popular stage the sense of "That Mesmerizing Mendelssohn Tune" revolves around the use of classical music by a black protagonist; the resulting satire is double edged, as the protagonist is mocked for pretensions to elite culture, which is itself mocked for the benefit of the popular theater's working class audiences through its appropriation by an unlettered protagonist. If the song were to be performed so that the identity of the protagonist is changed or obscured, its meaning would be different. Likewise the subversive message of an urban novelty or "suggestive" song depends on the audience's recognition that the protagonist, involved in some moral transgression, comes from a well-to-do urban background. This persona must be projected by the singer through appropriate diction, voice quality, and even stage deportment; any suggestion that the protagonist was from an ethnic or working-class background would negate the song's intended meaning. And the white female protagonists of Berlin's rhythmic/vernacular ballads assert their New Woman status through appropriation of sexual innuendos formerly found only in coon songs; any trace of racial dialect or hint of "black" singing style in their performance would negate this meaning.[88]

There are major problems with attempting to perform these songs today as they were done in the early twentieth century. At a practical level, the period recordings that might serve as models are not easily available: copies are preserved only in private or research collections, usually in delicate condition, and only a few have been reissued commercially.[89] And even if one succeeded in getting a copy of one song or another, ethnic and gender sensibilities have changed so much that a performance recapturing the original spirit and meaning might bring a storm of protest.

Though period recordings of these songs can be useful tools for scholars wishing to understand their original meanings, then, they are useful to performers not so much as precise models for emulation as of demonstrations of the flexibility with which performers of Berlin's own day treated his songs.

Public performance of popular music today is situated within a radically different tradition from that of the classical repertory. Rather than giving literal readings, performers (or their arrangers) are expected to shape any and all material to their own musical taste and personality, not merely through nuances of articulation and tempo but by flexible treatment of pitches and even formal structures. To put it another way, while the score of a piece of classical music is treated as a blueprint, with the visual symbols of musical notation transformed through a precise reading into sound, the score for a popular piece—if there is a score—is fleshed out according to the stylistic proclivities of the performer of the moment.

[88]The impact of performance on meaning and genre is explored in more detail in Charles Hamm, "Genre, Performance and Ideology in the Early Songs of Irving Berlin," *Popular Music* 13 (May 1994): 143–50.

[89]Four of Berlin's songs from this period, "Oh How That German Could Love," "Oh, That Beautiful Rag," "Woodman, Woodman, Spare That Tree!" and "Follow the Crowd," are included in the boxed CD set *Music From The New York Stage 1890–1920* (Pearl, Gemm CDS 9050-9061, 1993).

The third option for performers today follows from this: the score of any of these songs may be taken as a starting point for a new interpretation in which tune, text, harmony, and accompaniment are elaborated in a style characteristic of the performer, and the time,[90] in the grand tradition of popular music whereby both composer and performer contribute to the finished product. Berlin himself allowed for this when he recycled some of his early songs for use some decades after he wrote them.[91]

Summary

The foregoing discussion, analysis, and classification of Berlin's early songs suggest some conclusions:

- Berlin was not much interested—and also relatively unsuccessful according to the dictum that "a good song and a hit song are synonymous"[92]—in writing songs perpetuating the style and ideology of the genteel Victorian repertory, such as romantic, rustic, and domestic ballads.
- Even more than his peers, Berlin treated song genres and subgenres flexibly, often introducing elements of one into another for comic or dramatic effect and sometimes writing songs that almost defy classification.
- His earliest ethnic novelties are little different from those of other songwriters, but soon his protagonists, especially his black characters, depended less on the crude stereotyping typical of the nineteenth century and began taking on more universally human traits.
- He stood at the center of the development of a new subgenre, the urban novelty, that was accurately perceived in its day as breaking sharply with Victorian ideology and morality.
- His so-called ragtime songs have little to do with "classic" piano rags, but together with his later coon songs and his rhythmic/vernacular ballads they represent an early stage in the development of a generally syncopated style in which musical elements of African-American music are assimilated into the vocabulary of commercial urban songwriters. He was doing this as early as 1910, in ways that anticipated Tin Pan Alley styles of the 1920s and 1930s.
- Most of his early theater songs are merely songs of other genres, performed on the stage. Only at the end of this early period did he begin to confront the challenge of writing production numbers.
- Berlin played a considerable role in transforming the popular theater from a place shunned by the social and intellectual elite to one in which that group felt increasingly at home.

In sum, Berlin's least successful early songs are those in which he attempted to echo the stylistic and ideological heritage of America's past, and his best and most influential ones are those grounded in the radical urban multiculturalism of early twentieth-century New York City.

---

[90]As the Dutch singer Taco did in 1982 with Berlin's "Puttin' on the Ritz" and "Cheek to Cheek" (*After Eight*, RCA PL 28520), bringing them stylistically into the rock era. Berlin reportedly enjoyed these versions.

[91]For instance, "When the Midnight Choo-Choo Leaves for Alabam'," written as a coon song and performed as such on stage and recordings, was reorchestrated and stripped of all racial references when sung by Fred Astaire and July Garland in 1948 in the film *Easter Parade*.

[92]Wilder, *American Popular Song*, 92.

PLATE I. Cover for first edition of "Someone's Waiting for Me (We'll Wait, Wait, Wait)" (1909). Reproduced by permission of the Lester S. Levy Collection of Sheet Music, Special Collections, Milton S. Eisenhower Library, The Johns Hopkins University.

PLATE 2. Cover for first edition of "Next to Your Mother, Who Do You Love?" (1909). Reproduced by permission of the Lester S. Levy Collection of Sheet Music, Special Collections, Milton S. Eisenhower Library, The Johns Hopkins University.

PLATE 3. Cover for first edition of "Dreams, Just Dreams" (1910). Reproduced by permission of the Lester S. Levy Collection of Sheet Music, Special Collections, Milton S. Eisenhower Library, The Johns Hopkins University.

PLATE 4. Cover for first edition of "I'm Going on a Long Vacation" (1910). Reproduced by permission of the Lester S. Levy Collection of Sheet Music, Special Collections, Milton S. Eisenhower Library, The Johns Hopkins University.

IRVING BERLIN

EARLY SONGS

I. 1907–1911

I. MARIE FROM SUNNY ITALY

Words by
I. BERLIN

Music by
M. NICHOLSON

heart just yearns for you, dear, With fond af - fec - tion, And love that's
love, and say you're com - ing; The lit - tle birds, dear, Are sweet-ly
true, dear? Meet me while the Sum-mer moon is beam - ing,
hum - ming; Don't say, "No," my sweet I - tal - ian Beau - ty,
For you and me the lit - tle stars are gleam - ing;
There's not an - oth - er maid - en e'er could suit me,
Please come out to - night, my queen, Can't you hear my man - do - lin?
Come out, love, don't be a - fraid, Lis - ten to my ser - e - nade,
ritard.
ritard.

CHORUS
a tempo
p
p–f
My sweet Ma - rie from Sun-ny It-a - ly, Oh, how I do love you, Say that you'll
love me, love me, too, For-ev-er more I will be true, Just say the word and I will
mar-ry you, And then you'll sure-ly be, My sweet Ma - rie from Sun-ny It - a -
ly.
f
My sweet Ma -
ly.
1.
2.

2. QUEENIE

Words by
IRVING BERLIN

Music by
MAURICE ABRAHAM

18
Come out and meet me, la - dy, You are my dark-eyed ba - by,
Now, don't be hes - i - tat - ing, The birds a-bove are mat - ing,
22
Come, love, don't be a - fraid - y, My lit-tle hon - ey, do:
My love, I'll be re - lat - ing. To you my dusk - y queen:
26
Here while the stars am peep - ing And all the world am sleep - ing, My
Think, dear, of what you're miss - ing, Lov-ing and lots of kiss - ing, Come
30
prom - ise I'll be keep - ing, love, to you.
out, my love, and lis - ten to my plea.

34
CHORUS
Queen-ie, my own, I'll build a nice lit - tle home Where we can
mf
38
live all a - lone And from you I'll nev - er roam; Say you'll be mine,
43
Oh! hon - ey, please don't de - cline, And all the stars will then shine
47
for you and me.
1.
2.
me.
fz
D.S.

3. THE BEST OF FRIENDS MUST PART

Words and Music by
IRVING BERLIN

10
Where is your pay?
Been shoot - ing dice?
That shine a - bove,
That stuff don't go,
12
That's ver - y nice,
need - n't a - pol - o - gize,
You'd bet - ter blow,
can't buy a meal with that
14
I'm cold as ice,
Pack up your clothes,
Lord on - ly knows,
lov - in' you know,
I know it's hard,
On your old pard,
17
You've been the cause of my trou - bles and woes,
Don't ask me why,
But I can't help that you played your last card,
Don't hes - i - tate,

20
Too late to cry,
Just bid your-self good - bye.
E - vap - o - rate,
Hunt up an - oth - er mate.
22
CHORUS
A friend in need, is a friend in - deed, That's
p–f
25
just the kind of friend I've been to you, When
27
pov - er - ty smiled down on thee, I hung a - round and stuck like

30
glue, It's a long, long lane that has no turn, You can
33
nev - er tell your fin - ish when you start, But when you
35
find you can't make both ends meet, Then the
37
best of friends must part. A part.
1.
2.

4. I DIDN'T GO HOME AT ALL

Written and Composed by
EDGAR LESLIE and
IRVING BERLIN

9
night I have to talk some bus' - ness with some bus' - ness men, But
rath - er than dis - turb you from your slum - ber so di - vine, I
went to see a sick friend but had prom - ised to come back, At
11
rest as - sured that I'll re - turn at ten." That
lin - gered just to pass a - way the time, And
six - ty min - utes af - ter nine, to Jack. At
13
night he left the flat, and sim - ple wif - ey sat Up
then a - gain you see, I did - n't take the key, Which
six - ty min - utes past, next morn she came at last, Her

15
with the cat for Jack who nev - er came;
proves that my in - ten - tions were the best;
op' - ning speech was "I've been with my friend;
The
I
Be -
17
day was dawn - ing when he ram - bled home a - gain, And
e - ven said a pray'r, that you might not des - pair, Be -
cause they said she'd die, I want - ed to be nigh, And
19
said: "My dar - ling please let me ex - plain":
sides, I knew how much you need - ed rest":
so I wait - ed for the com - ing end":
20
CHORUS
I heard the clock strike One A. M., One A. M.,
p–f

24
One A. M., It told me of my prom - ise when I
27
left you in the hall; ___ Be - fore I knew 'twas Two A. M.,
31
Three A. M., Four A. M., But you know I prom - ised I'd be
34
home at ten, So I did - n't go home at all. ___ I all. ___
1.
2.
3
fz

5. DORANDO

Words and Music
by IRVING BERLIN

15
my pi - zon Pas - quale, He say we take da car And
one - a big - ga shame, I for - got da man's a - name Who
17
see Do - ran - do race a - "Long - a - ship;" Just
make me eat da I - rish beef - a stew; I
19
like da sport, I sell da bar - ber shop, And
ask - a him to give me da spa - gett, I
21
make da bet Do - ran - do he's a win. Then
know it make me run a - quick - a - quick, But I

23
to Ma-dees - a Square, Pas - quale and me go there, And
eat da beef - a stew, And now I tell - a you, Just
25
just - a like - a dat, da race be - gin.
like da pipps it make me ve - ry sick."
26
Allegro
Do - ran-do! Do - ran-do! He
CHORUS
run - a, run - a, run - a, run like
31
an - y-thing. One - a, two - a hun-dred times a - round da ring.

Slower
I cry, "Please - a nun - ga stop!"
Slower
Just then, Do - ran - do he's a drop! Good - bye,
poor old bar - ber shop. It's no fun to lose da mon, When de sun-of - a-gun no
a tempo
a tempo
run, Do - ran - do, He's good-a for not!
f
fz
Cym.
D.S.

6. NO ONE COULD DO IT LIKE MY FATHER!

Words by
IRVING BERLIN

Music by
TED SNYDER

15
fa - ther is the great - est man that ev - er wore a hat. He
had a cer - tain way to keep the la - dies in their place. Their
talk a - bout at - ten - tion, say! he gave her ev - 'ry glance. He
19
al - ways took things eas - y, in an eas - y sort of way, And
place was in the kit - chen, and his place to keep them there; I
has my moth - er danc - ing now, her brain is in a whirl, And
23
when it came to tak - ing things, just kind - ly let me say:
don't know how he did it, but I real - ly must de - clare,
on - ly here last week he came home with a string of pearls.
fz
27
CHORUS
No one could do it like my fa - ther! Ev - er clev - er,
p–f

4
My mother weighs three hundred pounds but don't give it away,
She bought a brand new sheath gown and she wore it yesterday,
My father showed his dignity when mother showed her sock,
Then just to make the two ends meet he used a big padlock.

CHORUS
No one could do it like my father
Ever clever, stunning, cunning father,
Neighbor Jones and his neighbor Lee
Are hunting for my father's key
And no one could do it like my dad.

5
I haven't told you how my father and my mother wed,
She was an old maid, he a burglar underneath her bed,
She flashed a gun at father and said, "I must be your wife,"
My father wears a medal now for saving someone's life.

CHORUS
No one could do it like my father
Ever clever, stunning, cunning father,
Mother thinks a lot of pa
But father drinks to think of ma
And no one could do it like my dad.

6
The other night when pa came home, he found to his surprise,
The iceman and my mother on the sofa making eyes,
He did not get excited, no! not one word did he say,
But when the iceman's bill came due, papa refused to pay.

CHORUS
No one could do it like my father
Ever clever, stunning, cunning father,
Father proved he was no slouch,
He fooled them all when he sold the couch
And no one could do it like my dad.

7
When father went to school they tell me he was very bad,
They also say he had a purpose to make teacher mad,
She'd make him stay in after school and pa would ne'er refuse,
For when it came to helping teacher tie her dainty shoes:

CHORUS
No one could do it like my father
Ever clever, stunning, cunning father,
Teacher knew a thing or two
She always wore a low cut shoe,
[And] no one could do it like my dad.

8
We lived right near a railway station not so far from here,
And father would make faces at the passing engineers,
They'd all throw coal at father, yes they would upon my soul,
And when the winter came around we never needed coal.

CHORUS
No one could do it like my father
Ever clever, stunning, cunning father,
First he got coal one by one
And now he sells it by the ton,
And no one could do it like my dad.

9
Around election time my father never knows his name,
Sometimes it's Breen, or Smith, or Green, Gilhouley or McShane,
Then other times it's Harrigan, O'Connor, or O'Dell,
They ought to call him "Winchester," 'cause he repeats so well.

CHORUS
No one could do it like my father
Ever clever, stunning, cunning father,
When you see him change his coat
You know that means another vote,
And no one could do it like my dad.

7. SADIE SALOME (GO HOME)

Words and Music by
EDGAR LESLIE and IRVING BERLIN

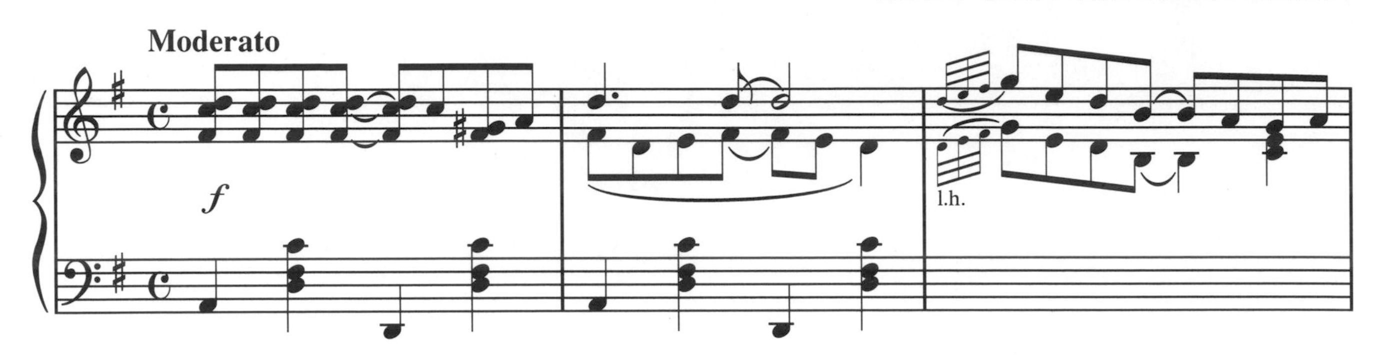

11
On the stage she soon be - came the rage, As the on - ly
You look sweet but jig - gle with your feet. Who put in your
14
real Sa - lo - my ba - by. When she came to town, her sweet-heart Mose
back such fun - ny mo - tions? As a sing - er you was al - ways fine!
17
Brought for her a - round a pret - ty rose; But he got an aw - ful fright
Sing to me, 'Be - cause the world is mine!' " Then the crowd be - gan to roar,
20
When his Sa - die came to sight, He stood up and yelled with all his might:
Sa - die did a new en - core, Mose got mad and yelled at her once more:

23
CHORUS
Don't do that dance, I tell you Sa - die,
p–f
25
That's not a bus' - ness for a la - dy!
27
'Most ev - 'ry - bod - y knows That I'm your lov - ing Mose,
29
Oy, Oy, Oy, Oy, Where is your clothes?

31
You bet - ter go and get your dress - es,
33
Ev - 'ry - one's got the op - 'ra glass - es.
35
Oy! such a sad dis - grace No one looks in your face;
37
Sa - die Sa - lome, go home.
1.
2.
home.
f
fz

8. MY WIFE'S GONE TO THE COUNTRY (HURRAH! HURRAH!)

Words by
IRVING BERLIN and GEO. WHITING

Music by
TED SNYDER

11
send me to the coun - try, dear, I know 'twould be a
Mis - ter Brown, come on down - town, I have some news for
14
treat." Next day his wife and fam' - ly were
you." He told a friend re - port - er just
17
seat - ed on a train, And when the train had
why he felt so gay, Next day an ad - ver -
20
start - ed, Brown - ie shout - ed this re - frain.
tise - ment in the pa - pers read this way.

CHORUS
My wife's gone to the coun-try, Hur-rah! Hur-rah! She
p–f
HUR-RAH!
HUR-RAH!
thought it best, I need a rest, That's why she went a-way, She
took the chil-dren with her, Hur-rah! Hur-rah! I don't care what be-
HUR-RAH!
HUR-RAH!
comes of me, My wife's gone a-way. My way.
1.
2.
D.S.

3
He sang his joyful story into a phonograph,
He made a dozen records, and I say it was to laugh,
For when his friends had vanished, and Brown was all alone,
His neighbors heard the same old tune on Brownie's graphophone.

CHORUS
My wife's gone to the country, hurrah, hurrah!
She thought it best, I need the rest, that's why she went away.
She took the children with her, hurrah, hurrah!
Like Eva Tanguay I don't care, my wife's gone away.

4
He went into the parlor and tore down from the wall,
A sign that read "God Bless Our Home" and threw it in the hall,
Another sign he painted and hung it up instead.
Next day the servant nearly fainted when these words she read.

CHORUS
My wife's gone to the country, hurrah, hurrah!
She thought it best, I need the rest, that's why she went away.
She took the children with her, hurrah, hurrah!
Now I'm with you, if you're with me, my wife's gone away.

5
He called on pretty Molly, a girl he used to know,
The servant said "She left the house about an hour ago,
But if you leave your name, sir, or write a little note,
I'll give it to her when she comes," and this is what he wrote.

CHORUS
My wife's gone to the country, hurrah, hurrah!
She thought it best, I need the rest, that's why she went away.
She took the children with her, hurrah, hurrah!
I love my wife, but oh! you kid, my wife's gone away.

6
He went and bought a parrot, a very clever bird,
The kind that always would repeat most anything she heard.
So when his voice grew husky, and Brownie couldn't talk
While he'd be taking cough-drops, he would have the parrot squawk.

CHORUS
My wife's gone to the country, hurrah, hurrah!
She thought it best, I need the rest, that's why she went away.
She took the children with her, hurrah, hurrah!
I knew my book, she left the cook, my wife's gone away.

9. JUST LIKE THE ROSE

Words by
IRVING BERLIN

Music by
AL. PIANTADOSI

17
knew no woe he loved her so, With
back to me so ten - der - ly, Would
21
love that a flow - er knows,
if they could on - ly last,
Un -
The
25
til one day from o'er the way, A
rose that sighed, and long since died, Seemed
29
hand plucked the lil - y fair,
fresh from the morn - ing dew,
The
Then

33
rose sighed and sighed, 'Til one day it died,
close to my breast, The sweet flow'r I pressed,
37
No - bod - y seemed to care.
And cried a - las for you.
41
CHORUS
Just like the rose dear I loved you, Like the lil - y they
mf
46
stole you a - way, Just like the rose I a -

51
dored you, Wor-shipped you night and day,
56
To - night all a - lone I am sigh - ing,
61
Sigh - ing for sweet re - pose;
Just like the
66
rose, my love's dy - ing, Dy - ing just like the rose.

10. OH, WHAT I KNOW ABOUT YOU

By JOS. H. McKEON
HARRY M. PIANO
W. RAYMOND WALKER

19
wee sis - ter Rose, Stood on her tip - toes And ea - ger - ly
her cun - ning way, He soon heard her say "My sis - ter said,
24
kept peep - ing through, Oh my, oh what bliss, She
she's not at home," Her face wore a grin, She
29
soon heard them kiss, As he gent - ly called her "my dear,"
said "Please come in," Poor Jo - seph oh my what he saw,
34
The ver - y next day When Flo came her
There sat pret - ty Flo, With her oth - er

38
way She whis - pered these words in her ear.
beau, Rose cried to her sis - ter once more.
43
CHORUS
Oh! what I know a - bout you,
Oh! what I
p–f
48
know a - bout you,
Don't say you did - n't 'Cause
53
I o - ver - heard,
Hon - est to good - ness I won't say a

58
word. Oh! what I know a - bout you,
63
Some-thing that I would-n't do,
Far be it from
68
me To re - peat what I see, But Oh! what I
72
know a - bout you.
1.
2.
you.
D.S.

11. SOMEONE'S WAITING FOR ME (WE'LL WAIT, WAIT, WAIT)

Words and Music by
EDGAR LESLIE and IRVING BERLIN

16
May, Soon they were call - ing their wives on the phone,
drinks, Man - hat - tan Cock - tails and High - balls ga - lore,
fe, They had grown tir - ed of wait - ing at home,

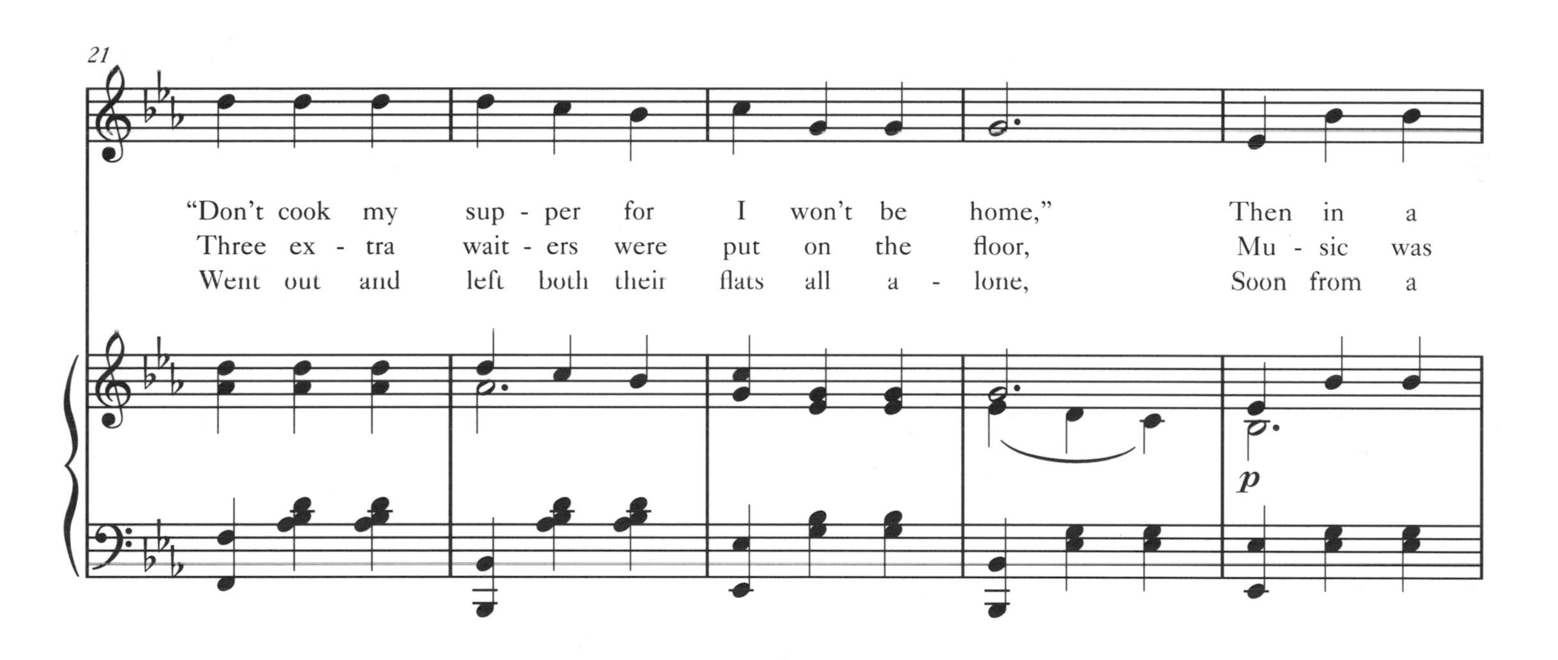
21
"Don't cook my sup - per for I won't be home," Then in a
Three ex - tra wait - ers were put on the floor, Mu - sic was
Went out and left both their flats all a - lone, Soon from a
p

26
"Tax - i" to some swell Ca - fe, They rode a - way to meet
play - ing "I'm com - ing home soon," O' Day was pray - ing for
cor - ner a voice hol - lered loud, "Give us an - oth - er and

31
Mol - ly and May, Wait - ed and wait - ed grew in - tox - i -
some oth - er tune, Hours were fly - ing, still they kept on
treat the whole crowd," Just then some - bod - y broke up the whole
36
cat - ed O' John - son then sang to O' Day.
buy - ing, O' John - son was cry - ing a - gain.
par - ty 'Twas Mol - ly, and sweet lit - tle May.
41
CHORUS
Some - one's wait - ing for me, Some - one's
f–p
46
wait - ing for you, I'm wait - ing and

51
you're wait - ing and wifey's wait - ing too,
56
Are we down-heart-ed, No!
We have
62
plen - ty of dough, WHAT! go home to wif - ey well, not on your
68
lif - ey we'll wait, Wait, Wait.
Wait.
1.
2.

12. DO YOUR DUTY DOCTOR!
(OH, OH, OH, OH, DOCTOR)

Words by
IRVING BERLIN

Music by
TED SNYDER

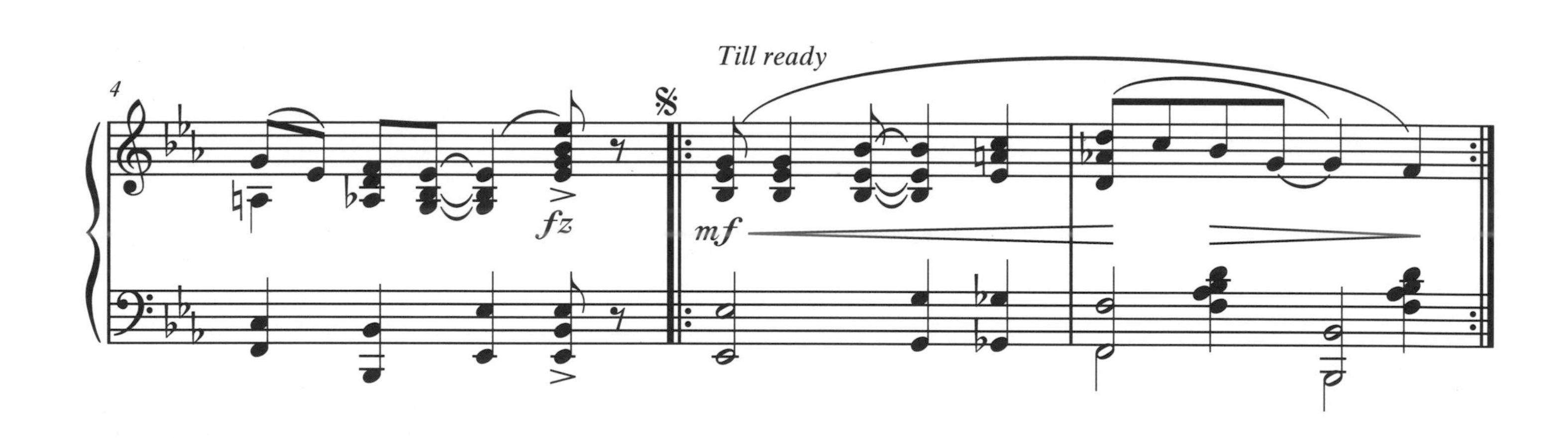

10
doc - tor quick, The doc - tor called a - round, these words to
is my line," Then start - ed lov - in' Li - za good and
13
say: "You're suf-f'ring from a
strong. Soon she was feel - ing
16
love at - tack, And if you want to
well once more, The doc - tor then looked
18
bring health back, A lov - ing man must love you ev - er - y
t'ward the door, And said, "I guess I'd best be get - ting a -

21
day." ___
long," ___
Then he turned to say good -
But E - li - za hol - lered
24
bye, Just to hear E - li - za cry: ___
quick, "Oh, I fear I'm get - ting sick." ___
fz
27
CHORUS
Oh, Oh, Oh, Oh, Doc - tor, ___ Won't you kind - ly hear my plea?
p–f
(Oh, doc - tor)
30
I know, you know doc - tor, ___ Ex -
(Oh, doc - tor)
(Oh, doc - tor)

33
act - ly what is best for me. ___
Hear me sigh, _
(Oh, doc - tor)
36
hear me cry, ___
Sure - ly you ___ ain't goin' to
38
let me die, _ For
if some love will
make me gain, _
ff
41
1.
2.
Do your du - ty doc - tor,
cure _ my pain.
cure _ my pain.
fz
D.S.

13. GOOD-BYE, GIRLIE, AND REMEMBER ME

Words by
IRVING BERLIN

Music by
GEORGE W. MEYER

14
girl - ie, To my own sweet - heart; It will
girl - ie, In a home for two; See just
17
wor - ry me, sweet girl - ie, And I'll be so lone - some
what it means, sweet girl - ie, What it means for you and
20
too, It will cheer me, dear, when you're not near me, Just to
I, So in June - time, spoon - time, hon - ey - moon-time, We'll be
23
know your heart beats true.
hap - py, bye and bye.

CHORUS
p–f
Good - bye girl - ie, and re - mem - ber me,
when you're far a - way,
I'll
be thinking-ing of you dear - ie, more and more each
day,
Sum - mer's com - ing, birds will

44
soon be hum - ming, love's sweet mel - o -
48
dy, Good luck!
52
good - bye, girl - ie, and re - mem - ber
56
1.
2.
me.
me.
f
fz

14. WILD CHERRIES

(Coony, Spoony Rag)

Words by
IRVING BERLIN

Music by
TED SNYDER

14
Jack-son was the lead-er of a big brass band,
Mis-ter Jack-son mar-ried Miss Lu-cin-da White,
17
Un-der-stand.
Sun-day night.
While at a ball one night,
Just as the knot was tied,
20
Lu-cin-da Mor-gan White,
Par-son had blessed the bride,
Yelled out with all her might,
Jack-son stood by her side,
22
When he start-ed play-ing some fa-mil-iar tune,
When the mu-sic played that lov-in' wed-din' march,

24
She just cried with de - light.
Miss Lu - cin - da just cried:
27
CHORUS
If you love yo' babe, play that tune for me,
mf
29
(Spoken)
Just be - cause I feel so fun - ny, Hon - ey I don't care for mon - ey,
No I'm not a blow - in' hon' but I just feel like go - in' some,
31
I'm goin' cra - zy that rag's a dai - sy I just can't make my

34
feel - ing be - have,
Play that tune _ a - gain,
Might - y soon _ a - gain,
37
(Spoken)
Hon - ey don't you start to scold, I'm feel - ing like a two year old,
Oh, you sev - en come e - lev - en! Hon I'm go - in' straight to heav'n
39
When I hear _ that strain, gwine to cry _ a - gain,
41
1.
2.
Play me some Wild _ Cher - ry Rag.
Rag. _
f
fz
D.S.

15. OH! WHERE IS MY WIFE TO-NIGHT?

By GEORGE WHITING,
IRVING BERLIN,
and TED SNYDER

Tempo di Valse

f

9

Till ready

mf *p*

You've heard a - bout Fox - y old Brown-ie, ___ Who
I've tried ev - 'ry way to raise mon - ey, ___ But

15

sent his whole fam' - ly a - way, ___ Well you can just bet that I
none of my friends are in town, ___ Ex - cept - ing the one man that

21

en - vied old Brown, When I heard him shout-ing Hoo - ray! ___ I
start - ed this thing, So I called a - round to see Brown, ___ He

27
bor - rowed some mon - ey from Brown - ie,
In - duced my old
an - swered, "I'm sor - ry old fel - low,
I'll have to re -
32
la - dy to pack,
It's now been a month and I
fuse you this time,
I just have e - nough mon - ey
37
can't raise the price, To pay for her fare com - ing back.
left in the bank, To send for that fam' - ly of mine."
42
CHORUS
Oh, where is my wife to - night,
Oh, why did she leave my

49
sight? I'm sor - ry to say, I shout-ed hoo - ray! oh, Why did I
56
ev - er let her go a - way, She nev - er would leave the town, If it
63
had - n't have been for Brown, Bring back, bring
marcato
70
1.
2.
back, Bring back my wif - ey to me. Oh, me.
f

16. SHE WAS A DEAR LITTLE GIRL

15
Same old case of, "I re-mem - ber you." Same old smile, and
Tho' it is - n't al - to-geth - er right, To re - mark a -
18
same old how - dy - do, Same old look of in - no - cence true, In her
bout her ap - pe - tite, Sev - en wait - ers worked hard that night, Serv - ing
21
great big eyes so blue.
what she called a bite.
CHORUS
She was a dear lit - tle
mf
3
24
girl, Dear - est of dear lit - tle girls,

27
Dear lit - tle eyes, dear lit - tle size, Dear lit - tle gold - en
30
curls, She mur-mured, "Dear nev - er fear,
33
I'll al - ways hold you dear." When she said so true, "I'm fond of
To a check his pen was in - tro-
36
wine," he knew, She was a dear, dear girl.
duced, but then,
1.
2.
girl.
f
3

17. SOME LITTLE SOMETHING ABOUT YOU

Words by
IRVING BERLIN

Music by
TED SNYDER

11
Girls who cap - ti - vate you with their rogu - ish smile,
I'll ad - mit it's made me jeal - ous as could be,
13
Girls you think be - yond com - pare,
That's be - cause I love you true,
15
When I saw the love - light in your eyes, sweet - heart,
Then a - gain, I thought of what the po - ets say,
17
When you smiled on me so ten - der - ly,
"Faint heart nev - er won a la - dy fair."

19
Mis - chief mak - ing Cu - pid pierced me with his dart,
So I don't in - tend to wait an - oth - er day,
21
Some - thing told me you were meant for me.
There's a co - zy home that we can share.
23
CHORUS
Some lit - tle some - thing a - bout you, Some lit - tle some - thing
p–f
26
dear, Means I can't live with - out you,

29
I al - ways want you near,
Some hid - den charm makes me
32
love you,
Some-thing I can - not see,
35
And I know one thing,
some lit - tle some - thing,
37
Means you're the girl for me.
1.
me.
2.
me.
fz

18. IF I THOUGHT YOU WOULDN'T TELL

Words by
IRVING BERLIN

Music by
TED SNYDER

Moderato con moto

f

4

Till ready

mf

A maid-en
One eve-ning

7

p

pret - ty, from Jer - sey Ci - ty, Paid a vis - it to some cou - sins out of
ear - ly, he took the girl - ie, To a res - tau-rant where drinks were of - ten

10

town; She an - swered, "Mer - cy" when cou - sin Per - cy, said, "I'd
tried; He or - dered por - ter, she or - dered wa - ter, Then she

13
dear - ly love to show you all a - round." Like bees a -
said, "I'll take a cock - tail on the side." She drank a
15
buzz - in' She and her cou - sin, Were to -
doz - en Just then her cou - sin whis - pered,
17
geth - er ev - 'ry min - ute of the day; One night he
"Dear - ie don't you think we should go home?" She an - swered,
19
kissed her, she shout - ed, "Mis - ter" But he
"La - ter, please call the wait - er," Then she

21
an - swered in a cute and cun - ning way.
mur - mured in a rath - er hus - ky tone.
22
CHORUS
If I thought you would - n't tell your moth - er, I would
p–f
25
try and take one more,
Prom - ise not to tell your
28
fa - ther or your broth - er, And I'll make it three or four,
If I

31
thought that you could keep a se - cret, I would
I would
33
keep you bus - y Nell,
take all they would sell,
I'd for -
35
get that I'm your cou - sin, And I'd take a half a doz - en,
37
1.
2.
If I on - ly thought you would-n't tell.
If I tell.
f

19. I WISH THAT YOU WAS MY GAL, MOLLY

Words by
IRVING BERLIN

Music by
TED SNYDER

9
looked so swell, I could - n't help but won - der, If a
did - n't mean to knock your oth - er sweet - heart, On the
11
gal like you would ev - er fall for me? I
lev - el, I am aw - ful sor - ry kid, I
13
know he thinks the world and all a - bout you, Well I
told me moth - er just how much I love you, It was
15
wish I had the chance to think so too, He
her that made me go to you and tell, So

17
would - n't be the same old boy with - out you, Gee, it
let me be your pal, if not your sweet - heart, When you
19
must feel great to have a gal like you.
mar - ry, I'll be there to wish you well.
20 CHORUS
I wish that you was my gal, Mol - ly, I
p
23
on - ly wish I was your beau, To

25
feel that you was my pal, Mol - ly, Would mean an aw - ful lot I
28
know, It is - n't right I know to tell you, But
31
what's a fel - low goin' to do? I wish that you was my gal,
34
Mol - ly, I on - ly wish you loved me too.
rit.

20. NEXT TO YOUR MOTHER, WHO DO YOU LOVE?

Words by
IRVING BERLIN

Music by
TED SNYDER

10
chil-dren in play - time, Two eyes of blue, gaz-ing in-to
cun-ning and pret - ty, Par-son had said, "You two are wed,
13
Eyes of her own sweet - heart,
Bless you my chil - dren, go."
One co-zy bench holds the
"Come spend a week," so she
16
cou - ple with - in it, Hours that passed on - ly
wrote to her moth - er, Moth - er re - plies, "I am
18
seemed like a min - ute, Sweet maid - en sighs,
bring - ing your broth - er." Hus - band re - marks,

20
Fel-low re-plies, "Tell me be-fore we part."
"Me for the parks. But first I want to know."
23 CHORUS
Next to your moth - er, who do you love?
p–f
25
Next to your broth - er, who do you love?
27
Next to your fa - ther, and your sis - ter Sue,

29
Tell me who looks good to you,
Next to your moth - er,
32
who would you miss?
Next to your broth - er,
who would you kiss?
35
poco a poco rall.
Next to your own dear folks at home,
poco a poco rall.
37
a tempo
Tell me, now who do you love?
love?
a tempo
1.
2.

21. STOP THAT RAG (KEEP ON PLAYING, HONEY)

Words by
IRVING BERLIN

Music by
TED SNYDER

13
Hon - ey, please be - lieve me, that my hear - in' ain't bad, But
Some - how or an - oth - er, hon,' my feet won't keep still, Please re -
15
some - thing seems to tell me that I'm list' - nin' to Rag; I've been sus -
quest that lead - er - man to keep a - play - in' till I cry e -
17
pect - in' so all night,
nough, e - nough some more;
Tunes I've heard be - fo' was on - ly
Hon' I'll ne'er for - give you, just for
20
mu - sic what ain't,
tak - in' me here,
Mel - o - dy like this is what an
Take me miles and miles a - way from

22
art - ist can't paint, Go get me some am - mo - nia, 'cause I
here an - y - where, Just raise a hand to move me, mis - ter
24
fear I must faint, I'm suf - fo - cat - ing with de - light.
man if you dare, I'm sat - is - fied to die right now.
26
CHORUS
Stop dat rag, keep on play - in' hon-ey, Stop dat rag,
p–f
30
hear me pray - in' hon - ey, Run a - way, I feel that I must be a - lone, But

33
don't you dare to leave me, 'till the cows — come home,
Cease dat strain,
fz
36
— please re - peat — it, hon - ey, Once a - gain,
and I'll meet it hon - ey,
39
Oh, I'm sad, — No, I'm — glad,
Stop dat rag, I'm say - in,' I
42
mean keep on a-play-in,' Don't you dare to stop dat rag. —
1.
2.
ff

22. CHRISTMAS-TIME SEEMS YEARS AND YEARS AWAY

Words by
IRVING BERLIN

Music by
TED SNYDER

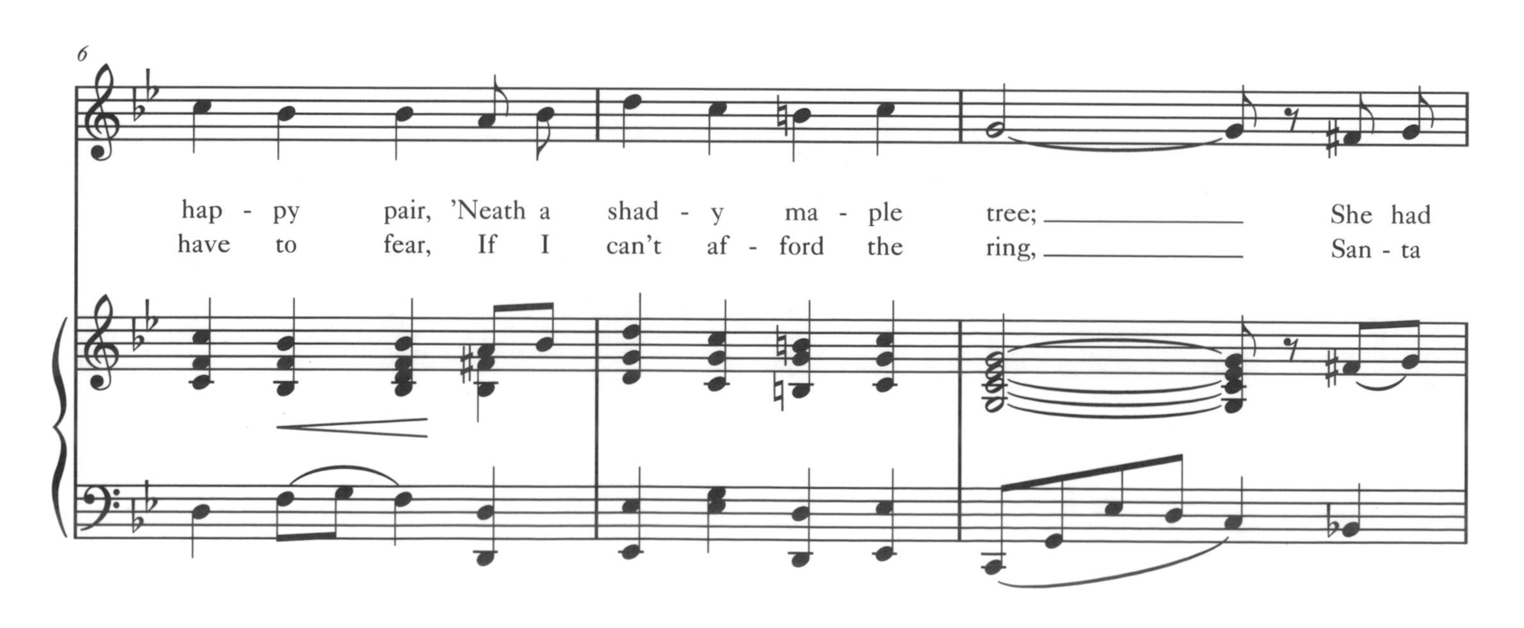

9
prom - ised him, "We'll be mar - ried, Jim, To the chimes of Trin - i -
Claus is kind, And I'm sure he'll find, Just a plain gold band to
12
ty, 'Tis the month of May, but next Christ-mas day, I will
bring, It's been just a year since I met you, dear, But it
15
be your blush - ing bride; Don't you wor - ry, dear, It will
seems just like a day, If I on - ly could, girl - ie
18
soon be here." But he looked at her and sighed.
dear, I would, Turn De - cem - ber in - to May."

22
CHORUS
Christ - mas chimes mean hap - py times, For
p–f
26
you and me, sweet - heart, Win - ter
31
time means you'll be mine, And nev - er more to
36
part, Hon - ey - moon can't

40
shine too soon, I wish it were to-
44
day, Girl - ie mine, don't
48
Christ - mas time, seem years and years a-
52
1.
2.
way?
way?
fz

23. YIDDLE, ON YOUR FIDDLE, PLAY SOME RAGTIME

Words and Music by
IRVING BERLIN

Moderato

mf *fz*

Till ready

p *p*

Ev - 'ry - one was sing - ing, danc - ing, spring - ing, At a wed - ding yes - ter - day, ___ Yid - dle, on his fid - dle played some rag - time, And when Sa - die heard him

At the sup - per ta - ble Sa - die thought, Yid - dle must have flew the coop; ___ She looked all a - round, but could not find him, 'Till she heard him drink - ing

14
play.
She jumped up and looked him in the eyes,
soup.
Sa - die wait - ed till they served the fish,
17
Yid - dle swelled his chest 'way out,
Ev - 'ry - one was
Then she jumped up - on the floor,
Put a quar - ter
20
tak - en by sur - prise, When they heard Sa - die shout.
right on Yid - dle's dish, And yelled to him once more.
23
CHORUS
Yid - dle in the mid - dle of your fid - dle, play some rag - time; Get
p–f

27
bus - y, I'm diz - zy, I'm feel-ing two years young, Mine choc-'late ba - by, if you'll
32
may - be play for Sa - die, Some more rag - time; Yid-dle, don't you stop, if you
36
do, I'll drop, For I just can't make my eyes shut up, Yid - dle on your
cresc.
fz
40
fid - dle, play some rag - time. rag - time.
1.
2.

24. I JUST CAME BACK TO SAY GOOD BYE

Words and Music by
IRVING BERLIN

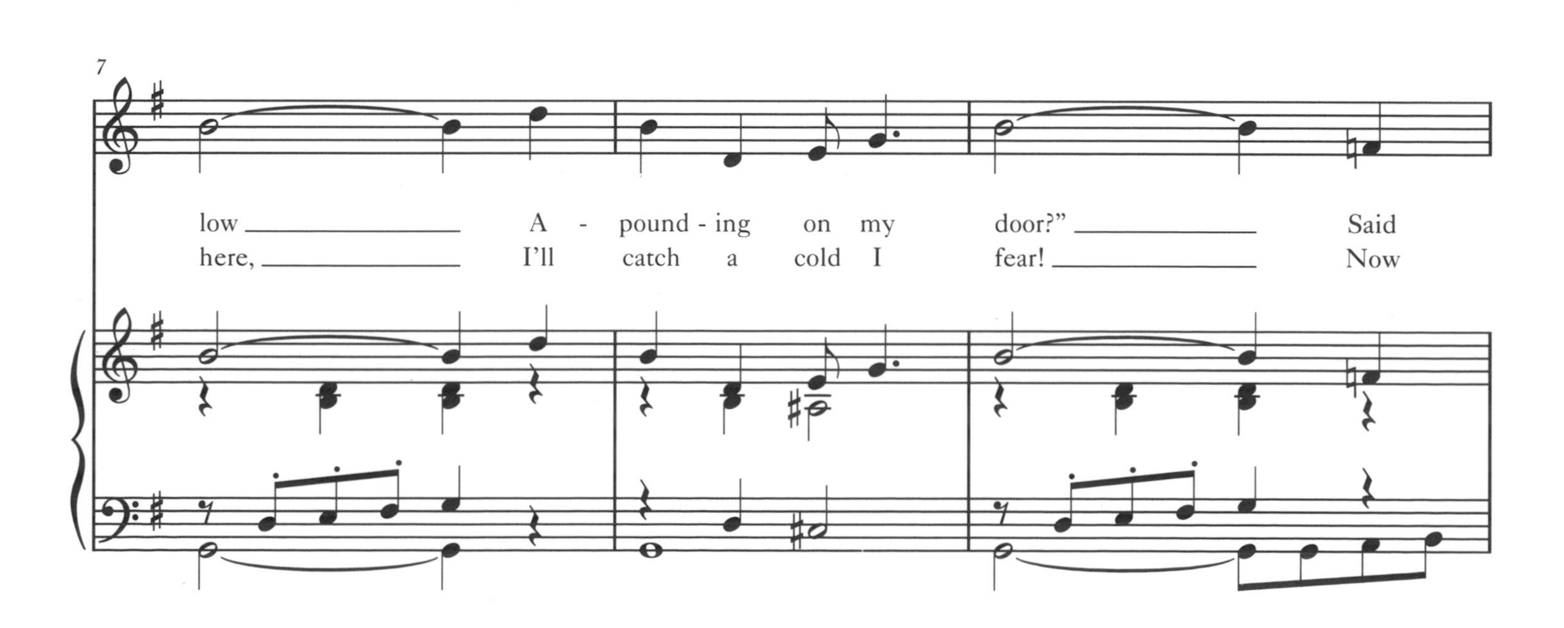

10
Li - za Green, "It's down-right mean To wake me from a
Li - za, please don't stand and tease! How can you see your
13
pleas - ant dream!" "It's Bill!" came the re - ply, "Came
Wil - liam freeze?" But Li - za said, "Go 'way! Come
16
back to say good - bye! I o - ver - looked it
'round some oth - er day." Then, as she closed the
19
when I left, That's just the rea - son why."
win - dow down, She heard poor Wil - liam say:
fz

CHORUS
I just came back to say, "Good-bye!" Let me ex-plain, I won't re-main; Be -
sides, you see, I don't know when I may have time to call a - gain. Been
cook - ing, hon - ey, I smell stew! Don't look so cross, I know you're boss! Just
tell me how you're feel-ing now, And then I'll say "Good - bye." I bye."
p– –f
f
fz
1.
2.

25. BEFORE I GO AND MARRY, I WILL HAVE A TALK WITH YOU

Words and Music by
IRVING BERLIN

9
Cry - ing 'cause her lit - tle heart was hurt;
That's the time I loved your ba - by smile;
11
Told her fel - low there was no de - ny - ing, She
When I saw how cute and neat was Dol - ly, I
13
saw him flirt. Fel - low whis - pers, "Let your heart be
liked your style. Fel - lows raved to me a - bout a
16
cheer - y, Won't you let me dry those tears a - way?
prize, dear, Mean - ing Car - rie's pret - ty eyes of blue,

19
Be your own sweet self and cud - dle near me,
So I went and gazed in - to her eyes, dear,
21
There is some - thing sweet I want to say."
That's the day I fell in love with you."
22
CHORUS
I may take a walk with Dol - ly, I may
p–f
25
smile at pret - ty May, I may have a talk with

28
Mol - ly, Or with Pol - ly, or e - ven Fay, I may
31
flirt with charm - ing Car - rie, As an - oth - er boy would
34
do; But be - fore I go and mar - ry, I will
37
1.
2.
have a talk with you. I may you.
fz
D.S.

26. THAT MESMERIZING MENDELSSOHN TUNE

Words and Music by
IRVING BERLIN

9
Won't you tell me how on earth you keep from sway - in'?
Is you gwine to wait un - til I'm al - most dy - in'?
11
ten.
Umm! Umm!
ten.
Oh, that Men-dels-sohn Spring Song
14
tune;
If you ev - er loved me show me now or nev - er,
Get your-self ac-quaint-ed with some real live woo - in',
17
ten.
Lord, I wish they'd play that mu - sic on for - ev - er,
Make some fun - ny nois - es like there's some - thing do - in',
Umm! Umm!
ten.

20
Oh, that Men-dels-sohn tune.
My hon - ey,
23
CHORUS Expressive and legato
Love me to that ev - er lov - in' Spring song mel - o -
p–f
26
dy,
Please me, hon - ey, squeeze me to that
stacc.
fz
28
Men - dels - sohn strain,
Kiss me like you would your moth - er,
ffz

30
One good kiss de - serves an - oth - er, That's the on - ly
32
mu - sic that was ev - er meant for me, That
35
tan - ta - liz - in', hyp - no - tiz - in', mes - mer - iz - in' Men - dels - sohn tune.
37
1.
2.

27. SOMEONE JUST LIKE YOU

Words by
IRVING BERLIN

Music by
TED SNYDER

11
Same game, still we must ex-plain, The old, old
Flo, Joe, on thro' life they go, In clo - ver
14
sto - ry must be told. Now to get their names, we'll call the fel - low Joe,
now the sto - ry's o - ver. Ev - 'ry song and sto - ry, has a hap - py two,
17
We could call him Har - ry, but Joe rhymes with Flo; Flo, Joe
Could-n't do with-out them, 'twould be some - thing new; Same tale
20
now their names we know, We'll have them say.
nev - er known to fail, When some - one sighs.

23
CHORUS
Some-one just like you, dear, Just like you will do, dear;
p–f
27
When I tell you true, dear, Can't you see?
31
Just your style, your size, dear, Just your smile, your eyes, dear;
35
1.
2.
Some-one just like you was meant for Some-one just like me. me.
f
fz

28. TELLING LIES

13
pale moon - light;
knot was tied;
Sal - ly said, "I think I'd bet - ter
Ev - 'ry morn he kiss - es her good-
3
16
go." As she rolled her eyes;
bye, Leaves her all a - lone;
But fox - y lit - tle Joe said,
But 'fore he goes a-way, These
20
"There's a game I know called tell - ing lies."
words he'll al - ways say, "I won't come home."
23
CHORUS
Tell - ing lies,
Tell - ing lies,
p–f

27
Say you nev - er want to be, Near to me, dear to me,
31
Tell - ing lies, close your eyes, Then a
35
kiss or two I'll steal and give them back to you, While
37
tell - ing lies.
1.
2.
lies.
f
ffz

29. SWEET MARIE, MAKE-A RAG-A-TIME DANCE WID ME

Words by
IRVING BERLIN

Music by
TED SNYDER

Moderato

mf

f

fz

Till ready

p

Liss - a to the sweet - a rag - a time, ______ Ain't it nun - ga make you feel - a fine? ______

What's-a mat - ta you no wan - na dance? ______ Nun - ga feel a - fraid to take a chance, ______

15
Ev - 'ry bod - y dance,
sleep - a like - a trance,
Ev - 'ry bod - y look,
shout - a, "Get da hook!"
17
Keep - a stead - y, get - a read - y,
One, two, three Kid!
I'm - a sad - a feel - a bad - a
Shake you - self Kid!
19
What's - a mat - ta you no look - a
wise?
If you love - a Ton - y nice - a
fine,
21
Get - a bus - y, make - a goo goo
eyes;
Make - a noise - a like - a rag - a
time;

23
Nun - ga make - a fake,
Hur - ry up - a quick,
just - a shake - a shake,
nun - ga take - a week,
25
Make - a jeal - ous for the oth - er guys.
If you nun - ga wan - na nev - er mind.
27
CHORUS
Sweet Ma - rie,
Make-a rag - a - time - a dance wid me?
p–f
30
Take a steam-a boat to It - a - ly,
On the rag - a line

33
Hey you wop - a, nun - ga stop - a kiss - a Ton - y call 'em pop - a
35
One, two, three I'm - a feel - a like - a "Hul - ly Gee,"
38
Come dance rag time wid me, Oh, my sweet - a Ma -
41
1.
rie.
2.
rie.
8-
fz
D.S.

30. IF THE MANAGERS ONLY THOUGHT THE SAME AS MOTHER

Words by
IRVING BERLIN

Music by
TED SNYDER

18
ac - tor folks, And lis - tened to their fun - ny jokes, So now I guess my
in that play, She said to me that ver - y day, "My child, she does - n't
22
moth - er ought to know; But man - ag - ers are aw - ful dense, It
know as much as you." So I got bus - y with my pen, And
fz
27
seems to me they have no sense, I know my dar - ling moth - er would - n't
wrote to Char - lie Froh - man then, I asked him if he would - n't star me
fz
31
lie, She thinks I'm bet - ter than the best; But man - ag - ers think
quick, His an - swer near - ly knocked me dead, For in his let - ter

36
I'm a pest, And all they say is, "Go some-where and die."
Char - lie said, "I did - n't ev - en know that you were sick."
40
CHORUS
If the man-ag-ers on - ly thought the same as moth - er, You
I'd
p–f
45
bet I'd show them all a thing or three, They'd have starred me in "The
act like Sa - rah Bern-hardt nev - er did, Har - ry Mil - ler would for -
50
Thief," For it's moth - er's firm be - lief; Mar - ga - ret Il - ling - ton could
get Mar - gret Ang - lin you bet, He would want me as his

54
nev - er steal like me.
real sup-port - ing kid.
If the Shu-berts on - ly took a tip from
If Froh-man on - ly took a tip from
59
moth - er,
I'd be star - ring like a star was nev - er starred;
moth - er,
He would la - bel me as Froh-man's on - ly star,
64
Les - lie Car - ter and that bunch Would be at Childs a - serv - ing lunch,
Mis - sus Fiske would take a drop To some Broad-way can - dy shop,
If the
69
man-ag-ers on - ly thought the same as ma.
1.
2.
ma.
mf
fz

31. OH HOW THAT GERMAN COULD LOVE

Words by
IRVING BERLIN

Music by
TED SNYDER

21
best what was walk - ing, well what's the use talk - ing, Was just made to
lov - ing was kill - ing, more yet she was will - ing, You nev - er would
well I re - mem - ber, one night in De - cem - ber, I felt like the
26
or - der for me. So love - ly, and wit - ty, more yet she was
have to say please. I just could - n't stop her, for din - ner and
mid - dle of May. I'll bet all I'm worth, that when she came on
32
pret - ty, You don't know un - til you have tried; She had such a
sup - per, Some kiss - es and hugs was the food; When she was - n't
earth, All the an - gels went out on pa - rade; No oth - er one
cresc.
38
fig - ure, it could - n't be big - ger, And there was some more yet be - side.
nice it was more bet - ter twice When she's bad she was bet - ter than good.
turned up, I think that they burned up, The pat - tern from which she was made.
fz

45
CHORUS
Oh how that Ger - man could love, With a
p–f
49
feel - ing that came from the heart, She
sweet - ness that's sweet - er than sweet, Just
love like you see in a play; When
53
called me her hon - ey, her an - gel her mon - ey, She
say what you please, you would hug and you'd squeeze, Just the
she said "My dear," it would ring in my ear, For a
57
pushed ev - 'ry word out so smart. She
shoes that she wore on her feet. Her
year, and a week and a day. Her

61
spoke like a speak - er, and oh what a speech, Like
smile was like mon - ey that some - bod - y owed you, That
no was like yes, and her yes was like no, It was
65
no oth - er speak - er could speak;
some - bod - y want - ed to give;
some - thing like yes, it was, well.
Ach my what a
When you felt like
When we got to -
70
Ger - man when she kissed her Her - man, It stayed on my
dy - ing and she start - ed sigh - ing, Ach my it was
geth - er ach don - ner und vet - ter, 'Twas love with a
74
1.
2.
cheek for a week. week.
worth-while to live. live.
cap - i - tal "L." "L."

32. WHEN YOU PLAY THAT PIANO, BILL!

Words by
IRVING BERLIN

Music by
TED SNYDER

15
round some Sun - day night, I'd love to have you play my
play the worst I know, While at my best to beat me
18
pia - no up - right." Said Mis - ter Wil - liam Brown, "The pleas - ure is mine, I can as -
so they must go," Then Miss E - li - za an - swered, "What's that you say? You on - ly
21
sure you la - dy fair." He called a - round next day, And start - ed
played the worst you knew! Sweet Wil - liam, I can guess how you play
24
in to play, Just to hear E - li - za de - clare.
at your best, But the worst right now will do."
fz

26
CHORUS
When I hear you play that pia - no so sweet, My
p–f

29
blood runs cold way down to my feet, You sure do bring forth

32
mu - sic Like I nev - er heard be - fore, When you

35
start in tear - ing rag by the streak, I could hear you play that

38
box for a week, For it does most an - y - thing but speak, When you

41
1.
2.
play that pi - a - no Bill. When I
fz

33. DRAGGY RAG

Words and Music by
IRVING BERLIN

14
feel - in' so fine, —
del - i - cate me, —
Won't you squeeze my hand a - gain? —
Hon - ey, dear, I'm suf-f'rin' so, —
17
Squeeze it hard, I feel no pain, —
Get your hat and we will go, —
Don't de - ny — me,
Don't de - cline, dear,
20
sat - is - fy — me to that drag - gy strain. —
take your time, — dear, let us walk out slow. —
fz
23 CHORUS
Oh! that drag-gy rag, —
Oh! that eas - y drag,
p–f

27
Don't you hur - ry, hon - ey, or you'll cause me pain, I
29
want to go to heav-en on an E - rie train,
Slide, glide,
32
not too fast,
Go slow to the last,
I love my quick de -
36
ci-sion, hon', But
Oh! that drag-gy rag.
1.
2.

34. DEAR MAYME, I LOVE YOU!

Words by
IRVING BERLIN

Music by
TED SNYDER

22
some - thing 'way down in my heart, I have - n't the nerve to
bar - gain, a nice wed - ding ring Worth twen - ty, I bought it for
27
poco rit.
say; But soon - er or lat - er I must make a
ten; Jim Jones wants to buy it for fif - teen next
poco rit.
32
rit.
start, So it might as well be to - day.
spring, Do you want me to sell it a - gain?
rit.
36
CHORUS Molto expressivo
I love you, I love you, I love you, Mayme,
p

44
I'd like to be there with a real pret - ty spiel, But three sim - ple
50
words can ex - plain how I feel, I love you, I want you, I
57
need you, is all that I know;
My love to your ma, kind re -
63
Più lento
gards to your pa, And kiss - es to you from Joe.
Più lento

35. GRIZZLY BEAR

Words by
IRVING BERLIN

Music by
GEORGE BOTSFORD

10
"Griz-zly Bear." All your oth - er lov - in' danc-es don't com - pare,
wait-er near, "Wa-ter, wa-ter quick the la-dy's gone, I fear."
13
Not so coon - y, But a lit - tle more than spoon - y,
"Thank you, hon - ey, In my purse you'll find some mon - ey,
15
Talk a - bout yo' bears that Ted - dy Roose-velt shot, They could - n't
Where's the man who show'd me how to do that dance, That put me
17
class with what old San Fran - cis - co's got.
in a trance? I'll take an - oth - er chance."

19
Lis - ten my hon - ey, do, and I will show to you the
Now that I've got my breath, I'm his - 'n un - til death, Come
21
dance of the griz - zly Bear.
on with yo' griz - zly Bear.
CHORUS
Hug up close to your
24
ba - by,
Throw your shoul-ders t'ward the ceil-in', Law - dy, Law-dy, what a feel - in',
Hyp - no - tize me like a wiz-ard, Shake yo'-self just like a bliz-zard,
27
Snug up close to your la - dy,
Close your eyes and do some nap-pin',
If they do that dance in heav-en,

30
Some - thing nice is gwine to hap - pen,
Shoot me hon' to - night at sev - en,
Hug up close to your
32
ba - by,
Sway me ev - e - ry - where,
35
Show your dar - lin' beau just how you go to Buf - fa - lo,
You and me is two, I'll make it one when we get thro',
37
Do - in' the Griz - zly Bear.
1.
2.
Bear.

36. CALL ME UP SOME RAINY AFTERNOON

Words and Music by
IRVING BERLIN

Moderato

mf

5

Nel - lie Green met Har - ry Lee,
He look'd wise, then looked for rain,

Till ready

p

p

11
He liked she, and she liked he, Just a case of love at sin - gle
"Give me three, four, five, six Main, Nel - lie dear, pre-pare I'm on my
14
sight;
way."
He took Nel - lie home that eve,
When he rang the front door bell,
17
Al - so took the num - ber of her phone,
No one there re-spond - ed to his call,
Just be - fore he
Soon he heard his
20
took his leave, Nel - lie whis - pered in the cut - est tone.
pret - ty Nell, Sing - ing to some-bod - y in the hall.

22
CHORUS
Call me up some rain - y af - ter - noon, I'll ar -
p–f
25
range for a qui - et lit - tle spoon, Think of
27
all the joy and bliss, We can
29
hug and we can talk a - bout the weath - er, We can

31
have a qui - et lit - tle talk, I will
33
see that my moth - er takes a walk, Mum's the
35
word when we meet, Be a ma - son, don't re - peat, An - gel
37
1.
2.
eyes are you wise? Good - bye. Call me bye.

37. THAT OPERA RAG

Words by
IRVING BERLIN

Music by
TED SNYDER

17
op - 'ra grand, Would nev - er fail to set this op - 'ra dark - ie
"Rig' - let - to," Sam said, "I know that's op - 'ra, be it here or
21
pp
wild." One eve - ning at a ball, They heard Sam
there." While he daubed up the wall He tried hard
pp
25
John-son call, Un - to the lead - er man to play some "Wil - liam
not to fall, But in the street the band kept right on play - in'
29
Tell"; The lead - er swelled his chest, and said, "I'll
fair, They played some "Faust" be - low, just then old

33
cresc.
do my best." So when he played they heard Sam John - son
Sam let go, He tum-bled down be-low a - shout-ing in the
cresc.
37
CHORUS
yell.
air.
Hear dat strain,
ff
p–f
40
Mis-ter Ver-di come to life a - gain, Oh that op-er-at - ic sweet re - frain,
44
Sho' would drive a cra-zy man in - sane, Just let me

48
die and meet those brain-y men, Who man-u-fac-tured
52
notes of op-er-a grand, Oh! Ver-di where, where oh
cresc
56
art thou? Let me shake you by the hand, man, you know what's grand,
60
Good Lord, it's ov-er, they're play-in' "Home Sweet Home."
ffz
p
ffz

38. I'M A HAPPY MARRIED MAN

Words by
IRVING BERLIN

Music by
TED SNYDER

15
sat - is - fied with bach - 'lor bliss, I took a wife you see." Said
would im - ag - ine mar - ried life was an - y - thing but fine; They

19
Gray to Smith, "You're cra - zy man, why, mar - ried life is fine, I
don't ap - pre - ci - ate the wife who helps to make life gay, But

23
have the dear - est lit - tle wife and let me say for mine."
far be it from me to frown, I mean it when I say.
fz

27
CHORUS (Slower)
I'm a hap - py mar - ried man, Yes, I am, yes, I am,
(Slower)
p–f marcato
31
Mar - ried life was made for me, I would - n't choose an - oth - er;
35
Betch yer life I love my wife, love my wife, with all my life,
39
1.
2.
I'm a hap - py mar - ried man, my wife lives with her moth - er. moth - er.

39. I LOVE YOU MORE EACH DAY

Words by
IRVING BERLIN

Music by
TED SNYDER

22
for my prize; The hand of time has turned your gold - en
in - to day; You nev - er mea - sured what you gave by
29
locks to sil - v'ry gray, The sil - ver threads have
what you thought I'd give, You're still my lit - tle
35
drawn you clos - er to my heart each day.
sweet - heart, and I'll love you while I live.
40
CHORUS
You were my queen at sweet six - teen, You're my queen at six - ty
p–f

47
three, Your eyes that shone with glad - ness then, Still hold their
54
charms for me, And as I gaze in - to your eyes, there's
61
one thing I must say, You're still the same sweet dar - ling
68
1.
2.
girl, And I love you more each day. You day.

40. ALEXANDER AND HIS CLARINET

BERLIN and SNYDER

15
Miss E - li - za John - son was his an - gel pet,
When he got in - side he played and played like sin,
17
And Al - ex - an - der was her one best bet;
Then played her cards to see who'd buy the gin;
19
Strange to say they quar - reled on last Sun - day night,
When he left, Miss Li - za tried some sleep to get,
21
Mon - day eve - ning Al - ex - an - der came in sight,
Dreamt her Ro - me - o came back to Ju - li - et,

23
Played his clar - i - net be - neath her win - dow light,
Al - so dreamt he brought with him his clar - i - net,
25
To hear E - li - za yell with all her might.
If no - one woke her, she'd be shout - in' yet.
27
CHORUS
Hon - ey, is that you? yes, yes,
p–f
30
Did - n't ev - en have to guess, my hon - ey, what brought you?

33
Oh pet, I see you brought your clar - i - net, My hon - ey,
36
I'm an - gry, no, no, For law - dy sake don't dare to go, My
39
pet, I love you yet, And then be -
41
sides, I love your clar - i - net.
1.
2.
net.

41. SWEET ITALIAN LOVE

Words by
IRVING BERLIN

Music by
TED SNYDER

9
Call - a da sweet - a name - a like - a da dove,
There's one - a man up in da moon all - a right,
11
It make me sick - a when they start in to speak - a 'Bout the
But he no tell - a that some oth - er nice fel - ler Was - a
13
moon 'way up a - bove.
kiss your gal last night.
What's-a da use to have - a
May - be you give your gal da
16
big - a da moon?
wed - ding - a ring,
What's - a da use to call - a
May - be you mar - ry, like - a

18
dove?
me,
If he no like - a she,
May - be you love your wife,
20
and she no like - a he, The moon can't make them love; But,
may - be for all your life, But dat's on - ly may - be. But,
23
CHORUS
Sweet I - tal - i - an love, Nice I - tal - i - an love,
p–f
27
You don't need the moon - a - light your love to tell her,
When you squeeze your gal and she no say, "Please stop - a!"

29
In da house or on da roof or in da cel - lar,
When you got dat twen - ty kids what call you, "Pa - pa!"
Dat's I -
32
tal - i - an love,
Sweet I - tal - i - an love;
When you
35
kiss - a your pet And it's - a like - a spa - gette,
kiss one - a time And it's - a feel like - a nine,
ff
37
1.
2.
Dat's I - tal - i - an love!
love!
fz

42. OH, THAT BEAUTIFUL RAG

Words by
IRVING BERLIN

Music by
TED SNYDER

15
Just nev - er mind the name, Rag - time is all the same,
Bring on yo' bill o' fare, Hon - ey, I do de - clare,
17
Mu - sic is mu - sic with me.
But I will say that it's
Some-how I'm feel - in' for - lorn.
Hear them play - in' that old
20
beau - ti - ful, hon', With a great beg cap - i - tal B.
beau - ti - ful rag, Now my ap - pe - tite is gone.
23
CHORUS
Oh! oh! oh! oh! Oh! that beau - ti - ful rag,
It sets my heart a - reel - in,'

27
Oh! oh! oh! oh! Oh! that beau - ti - ful drag,
30
That fun - ny feel - ing steal - ing, Hear that trom - bone blow - in', hon',
33
Ain't dem fid - dles go - in' some? Oh! sir, Oh! sir, cud - dle up clos - er,
37
Squeeze me like you would a flow - er, Make a min - ute last an hour, Oh! oh! oh! oh!

40
Oh! that heav - en - ly strain, It makes me feel so fun-ny; If I ev - er
44
cry, "Don't play it a - gain," Just don't be-lieve me, hon-ey, Oh, my dear - ie,
48
can't you hear me call-in'? Come up near me, catch me, dear, I'm fall-in', Oh! oh! oh! oh!
52
1.
2.
Oh! that beau - ti-ful rag. rag.

43. TRY IT ON YOUR PIANO

Words and Music by
IRVING BERLIN

11
pi - a - no all the time, Like a reg' - lar Ru - bin -
cures your pain and care, He was known most ev' - ry -
14
stein. Sun - day he called a - round to
where. Lu - cy took sick one day, he
16
see Miss Lu - cy Brown And said "My dar - ling
called a - round to say "I've brought with me a
18
pet, I have found a new way to make love That
pill, It's a new dis - cov - 'ry of my own That

21
has - n't been dis - cov - ered
sure - ly ought to cure or
yet,
kill,
Won't you let me show you
It has nev - er yet been
fz
24
how?"
tried,"
But Miss Lu - cy cried "not
But Miss Lu - cy loud - ly
now":
cried:
fz
27
CHORUS
Try it on your pi - a - no grand,
I don't care to
p–f
30
un - der - stand
B or I flat,
C or Y flat,

33
Try it hon' but not in my flat.
While I don't doubt that what you
Give me Per u - na for my
36
say is true, I'm not tak - ing chanc - es with some
ev - 'ry pain, For he who takes that will live to
38
love that's new, So Mis - ter Man - ner, try it
take a - gain, So try your brand up on a
40
on your pi - a - no, But you can't try it on me. me.
ba - by grand, Be-cause you can't try it on me. me.
1.
2.
f
fz
8-

44. "THANK YOU, KIND SIR!" SAID SHE

Words by
IRVING BERLIN

Music by
TED SNYDER

15
Some - one's fall - en down, slipped up - on the ground, It hap - pens on a slip - p'ry
Maid - en soft - ly speaks, "I've been sad for weeks, My dar - ling hus - band failed last
18
street.
year.
Kind young gent, who saw the ac - ci - dent,
I'm in debt right up to here, you bet."
21
poco rit.
Helps the charm - ing la - dy to her feet.
Fel - low whis - pers some - thing in her ear.
poco rit.
23
CHORUS
"Thank you, kind sir," said she. "Wel - come, sweet Miss," said he.
p–f

27
"May I not es - cort you home?"
"You may not, I'll go a - lone,
I would hate to go a - lone."
molto cresc.
31
I've a hus - band," said she, "And I've a wife, Miss," said he, "You
"I live here, sir," said she, "And so do I, Miss," said he. "Don't
p
f
p
p
f
p
35
need - n't call me up," said she, "My 'phone is one, two, three."
dare to call on me," said she, "I'm al - ways in at three."
f
37
1.
2.
"Thank you," said he.
he.
p
fz
D.S.

45. YIDDISHA EYES

Words and Music by
IRVING BERLIN

11
Jen - ny Gold - en Dol - lars was there. Now comes the bus' - ness end,
Jen - ny's bus' - ness fa - ther got wise! Now comes the fin - ish part,
14
Ben - ny, to a friend, said, "In - tro - duce me to Miss Jen - ny, do!"
Ben - ny, from his heart, be - gan to shout, "I love your daugh-ter fine!"
17
Soon he whis - pered, "Pleased to meet you too!" Then young Ben - ny
Jen - ny's pa cried, "That's how I got mine!" Ben - ny, bend - ing,
20
winked at Jen - ny With his Yid - dish - a eyes.
saw the end - ing, With his Yid - dish - a eyes.

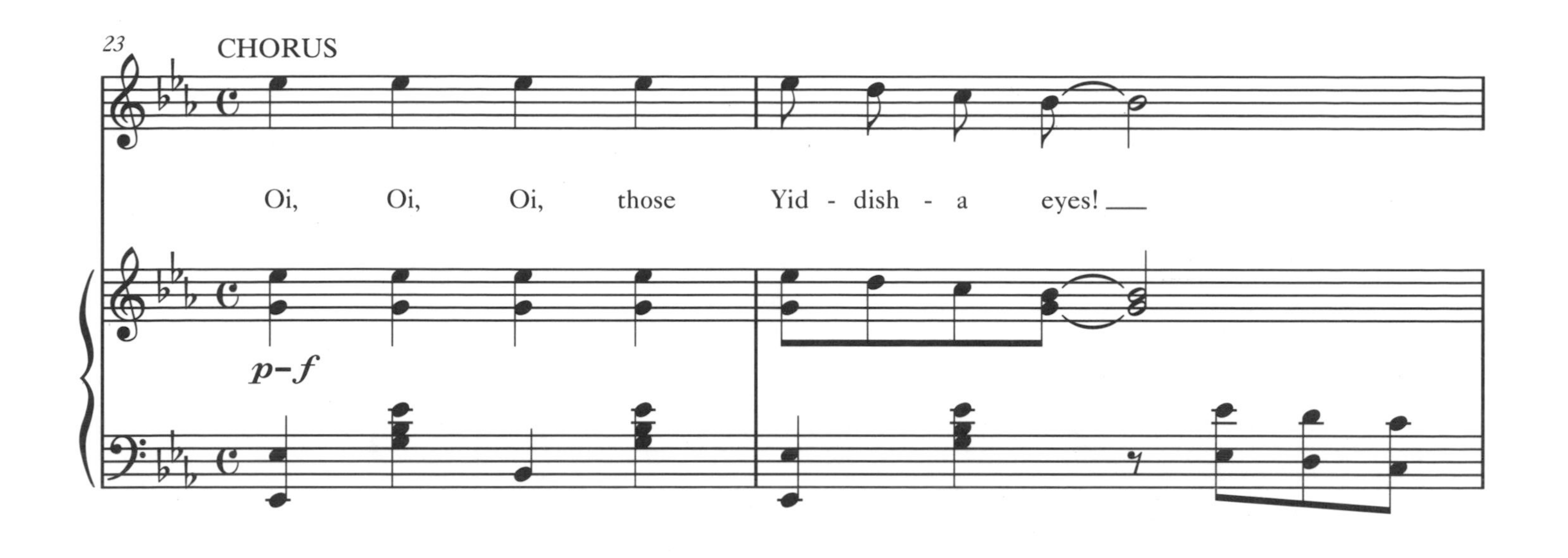
23
CHORUS
Oi, Oi, Oi, those Yid - dish - a eyes!
p–f

25
Ben - ny had those Yid - dish - a eyes, That shone so bright, with an

28
Is - ra - el light; Eyes that could tell a dia-mond in the night.

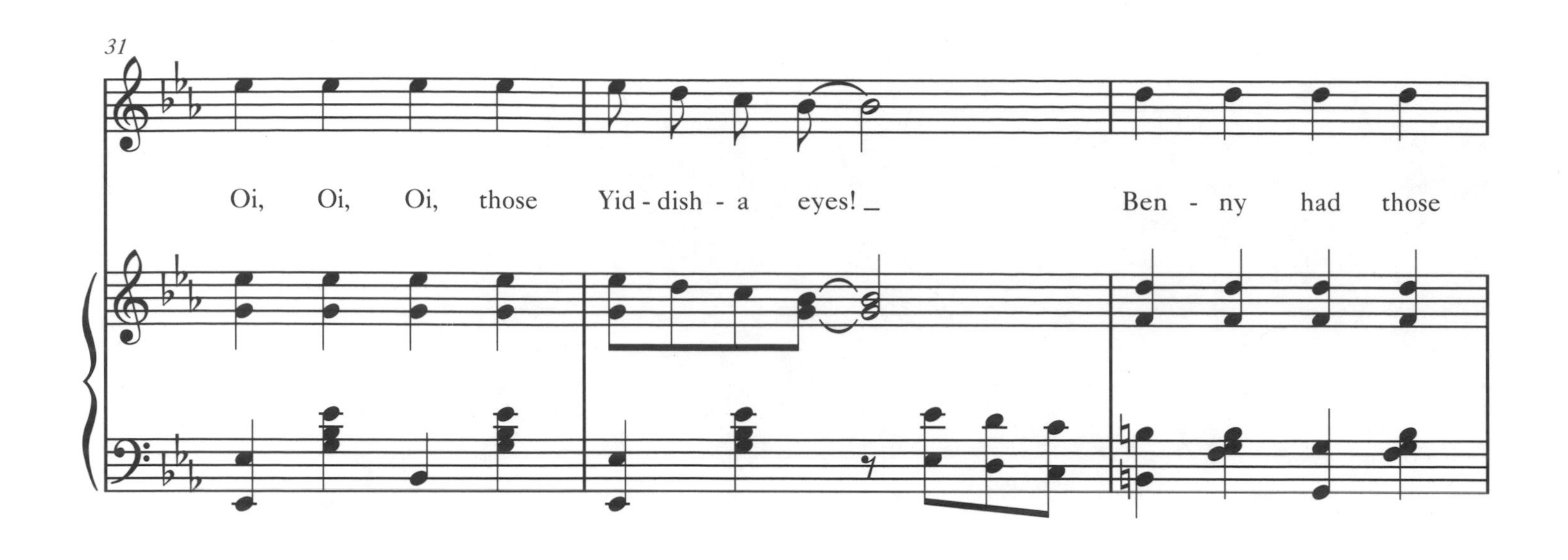
31
Oi, Oi, Oi, those Yid - dish - a eyes! _ Ben - ny had those

34
Yid - dish - a eyes! _
He took a look in her bank - book,
He said, "Fare - well," the tear - drops fell

37
1.
2.
With his Yid - dish - a eyes. _____ eyes. _____
From his Yid - dish - a eyes. _____ eyes. _____

46. IS THERE ANYTHING ELSE I CAN DO FOR YOU?

Words by
IRVING BERLIN

Music by
TED SNYDER

11
Wash-ing the dish - es, feed-ing the fish - es, As ev-'ry girl should
There was no stop - ping, John start-ed chop - ping, Chopp'd for an hour or
14
do.
two.
John-ny dropped in and soon made him-self hand - y,
Soon work and John - ny be - came per-fect strang - ers,
17
Help-ing his Ma - bel dear,
He did the best he could,
When all the work was thro',
Call-ing his Ma - bel near,
20
and noth-ing more to do, He whis-pered in her ear.
he whis-pered, "Lis - ten, dear, I'm not so strong for wood!"

22
CHORUS
Is there an - y - thing else that I can do for you?
p–f
25
Tell me do, come on and ask me to, I've got
27
hours to spare, my dear, and then a few,
29
Tell me Ma - bel, I'm a - ble to do

31
an - y - thing else, so don't you be a - fraid.
33
Pret - ty maid, I've learned the tai - lor trade, And some -
35
how I've been guess - in', That your lips need some press - in', Tell me,
37
dear, is there an - y - thing else? Is there else?
1.
2.

47. KISS ME MY HONEY, KISS ME

Words by
IRVING BERLIN

Music by
TED SNYDER

12
CHORUS
Kiss me, my hon - ey, kiss me, And say you'll
p–f delicato
16
miss me as I'll miss you;
20
Love me, my hon - ey, love me, Like stars a -
24
bove me, say you'll be true while a-way ev-'ry day, I'll be think-ing of you,

28
Dear - ie, now don't grow wea - ry, Be bright and
32
cheer - y, my hon-ey do, So dear,
37
be - fore I go, dear, Come here and kiss me, (Kiss, Kiss.)
41
hon-ey, I love you.
1.
2.
you.
rit. e cresc.
fz

48. COLORED ROMEO

Words by
IRVING BERLIN

Music by
TED SNYDER

12
Then hunt - ed for a man;
Soon she was
They ran off in a car;
They made a
p
15
dream-ing of
A beau who cooed just like a dove,
mile or three
In al - most noth - ing, don't you see,
18
A man who sho' could love
Like a Rom - e - o
The au - to struck a tree
Then they did - n't go
21
can.
She wan - dered out one night
far.
Miss Li - za from the ground

24
— be - neath the pale moon - light And sang with
— kept gaz - ing all a - round Her beau could
27
all her might This an - cient cry.
not be found So Li - za cried.
CHORUS
30
Where art thou Ro - me - o, My col - ored
p–f
33
Ro - me - o, The bal - co - ny wants to see he and

36
she, you and me come out and pet your col - ored Ju - li -
38
et. Don't lin - ger do it now, Come kiss my
41
fe - vered brow, Shake - spear's love come and show me,
44
1.
2.
col - ored Ro - me - o. Where art thou o.

49. STOP, STOP, STOP
(COME OVER AND LOVE ME SOME MORE)

Words and Music by
IRVING BERLIN

Moderato

mf

5

Till ready

p

Hon - ey, there's some - thing buzz-in'
Hon' did I hear you say you're

8

'round my heart, Some-thing that must be sat - is - fied, My dear - ie,
go - ing home? Just 'cause the clock is strik-ing nine, My dear - ie,

11

See that Mor - ris chair Stand - ing o - ver there,
That clock at ___ its best, Is an hour ___ fast,

13
There's some room to spare, Now for some love pre-pare.
Eight o'-clock just past, Stay, let the par-ty last.
15
Make your-self comf'-ta-ble be-fore we start,
Sure-ly you would-n't leave me all a-lone,
17
Tie your-self right up to my side,
Just for to sat-is-fy the time,
Sing me that lov-in' song
20
that goes some-thing like
Lento
p
umm
umm, umm, umm.
Lento

CHORUS
a tempo
Cud - dle and squeeze _ me hon - ey,
a tempo
p–f
Lead me right to cu - pid's door, ___ Take me
out up - on that o - cean called the "Lov - a - ble Sea," ___
Fry each kiss in hon - ey, then pre - sent it to me. ___

31
Cud - dle and please me hon - ey,
33
An - chor at that kiss - ing shore; My hon - ey
35
stop, stop, stop, stop, don't dare to stop, Come
37
o - ver and love me some more.
1.
2.
more.

50. HERMAN LET'S DANCE THAT BEAUTIFUL WALTZ

Words by
IRVING BERLIN

Music by
TED SNYDER

19
stood Mis - ter Her - man, a sweet lit - tle Ger - man, The
ver - y same feel - ing I feel on me steal - ing And
23
best that was dressed in the hall.
Her - man I'm look - ing at you.
The mu - sic played
So close both your
28
danc - es like rag - time and lanc - iers, But no one would go on the
eyes make be - lieve you ain't wise, On - ly puck - er your lips in - to
33
floor,
place,
So quick - ly the band played a waltz that was
Think of five hun - dred meld or a sweet An - na
cresc.

38
grand, A waltz that made sweet Le - na roar.
Held, While I kiss the hole in your face."
43
CHORUS
Her - man let's dance to the
47
tune of that beau - ti - ful waltz, Now lis - ten you
52
Ger - man, I'm talk - ing to you, I'll do some - thing dear you don't

57
want me to do, so come on take a chance
62
and I'll know that your love is - n't false. A
67
feel - ing that's heal - ing, comes steal - ing while spiel - ing that
71
beau - ti - ful, beau - ti - ful waltz.
1.
2.
waltz.

51. PIANO MAN

By BERLIN and SNYDER

13
No? then you've yet to live. He sits on his stool like a
rain - y af - ter - noon. They all need - ed love to in -
16
king on the throne, And plays and plays with ease, Why the
spire their notes, Be it the spring or fall, But my
19
mel - o - dy just nes - tles in his fin - ger tips And
pia - no man, just have him meet a pia - no grand Then
21
oo - zes out in the keys. Pi -
lis - ten, Umm, Umm, that's all. Pi -

23
CHORUS
a - no man, Pi - a - no man, He brings forth notes like
p–f
26
no one can, Oh what a feel - in' When his
L.H.
R.H.
28
notes come a steal - in', why I just feel like kneel - in' and ap -
30
peal - in' to my Pi - a - no man, Pi - a - no man,

33
Law - dy how his mu - sic lin - gers, May the Heav - en bless his fin - gers.
35
When he plays for days and days, It soothes me like a
38
fan. Just lend your ear, dear, here, near to my
stacc.
41
1.
2.
ev - er lov - in' pia - no man. Pi - man.

52. INNOCENT BESSIE BROWN

Words and Music by
IRVING BERLIN

11
Fel-lows looked at her say-ing what a pit - y, She's from the farm
Fel-lows would call up - on her by the doz - en, In - no-cent Bess,
14
mean-ing no harm, Ev - 'ry maid - en, ev - 'ry gent,
nev - er could guess, Where her cous - in got such toys,
17
When dis - cuss - ing she, Said she must be
Dia - monds more or less, 'Twas - n't long be -
20
in - no - cent, 'Cause she came from Kan - ka - kee.
fore the boys Made a jew'l - ry store of Bess.
fz

23
CHORUS
In - no - cent Bes - sie Brown Want - ed to see the
In - no - cent Bes - sie Brown Dressed in a coun - try
p–f
26
town, Met up - on the Av - e - nue one
gown, Bes - sie Brown was lone - some for a
28
af - ter - noon fair, A fel - low who said "'ave - n't you an
dia - mond or two, But since she met the cit - y she could
30
hour to spare?" So pret - ty In - no - cent Bes - sie
loan some to you, And she was In - no - cent Bes - sie

32
Brown, Went for an hour in town, And when the
Brown, Lived in a coun - try town, When Mis - sus
35
day was break - ing they were still par - tak - ing Of some
Mor - gan chub - by, missed her dar - ling hub - by, She be -
37
fizz - es of gin that fizzed with - in,
gan talk - ing war while hunt - ing for
39
1.
2.
In - no - cent Bes - sie Brown.
Brown.
f
fz
D.S.

53. DREAMS, JUST DREAMS

heart For now that we've drift - ed a - part, I find joys a - new while
seems, Where night nev - er turns to sun - beams, The world I would give to
dream - ing of you, Dreams that can ne'er come true.
go there and live, Live with my gold - en dreams.
rit.
CHORUS con espressione
Dreams, just dreams, My beau - ti - ful gold - en
dreams, It seems my dreams Are mis-sives of sweet con - so -
poco rit.

28
a tempo
la - tion. Each dream I dream Turns gloom to a bright sun -
a tempo
mf
32
rit. e cresc.
beam, Since the love that I gave Found a grave, all I crave Is
rit. e cresc.
35
a tempo after 1st verse
dreams, just dreams.
a tempo
D.C.
39
a tempo after 2nd verse
dreams, Just dreams.
a tempo
allargando
f

54. I'M GOING ON A LONG VACATION

Words by
IRVING BERLIN

Music by
TED SNYDER

16
la - dy thin, Calls a - round, is ush - ered in,
where she'd go, I can tell your wife you know,
19
sees her hus - band fond - ly kiss Miss Nel - lie, Next
Don't you think I bet - ter go with Nel - lie," Next
23
morn - ing Kel - ly whis - pered in Nell's ear, She
week the of - fice boy and Nel - lie Brown, Were
27
left the of - fice sing - ing loud and clear.
hold - ing hands and sing - ing out of town.

CHORUS
Slowly
I'm go-ing on a long va - ca - tion, Oh you rail - way sta - tion,
p–f
First in years so give three cheers yea - bo, yea, hur - ray hur - ray,
Ev-'ry-bod-y in cre - a - tion, Needs some rec - re - a - tion,
My boss guessed I need some rest, yea, I'm go-ing a - way. I'm way.
1.
2.

55. BRING BACK MY LENA TO ME

By BERLIN and SNYDER

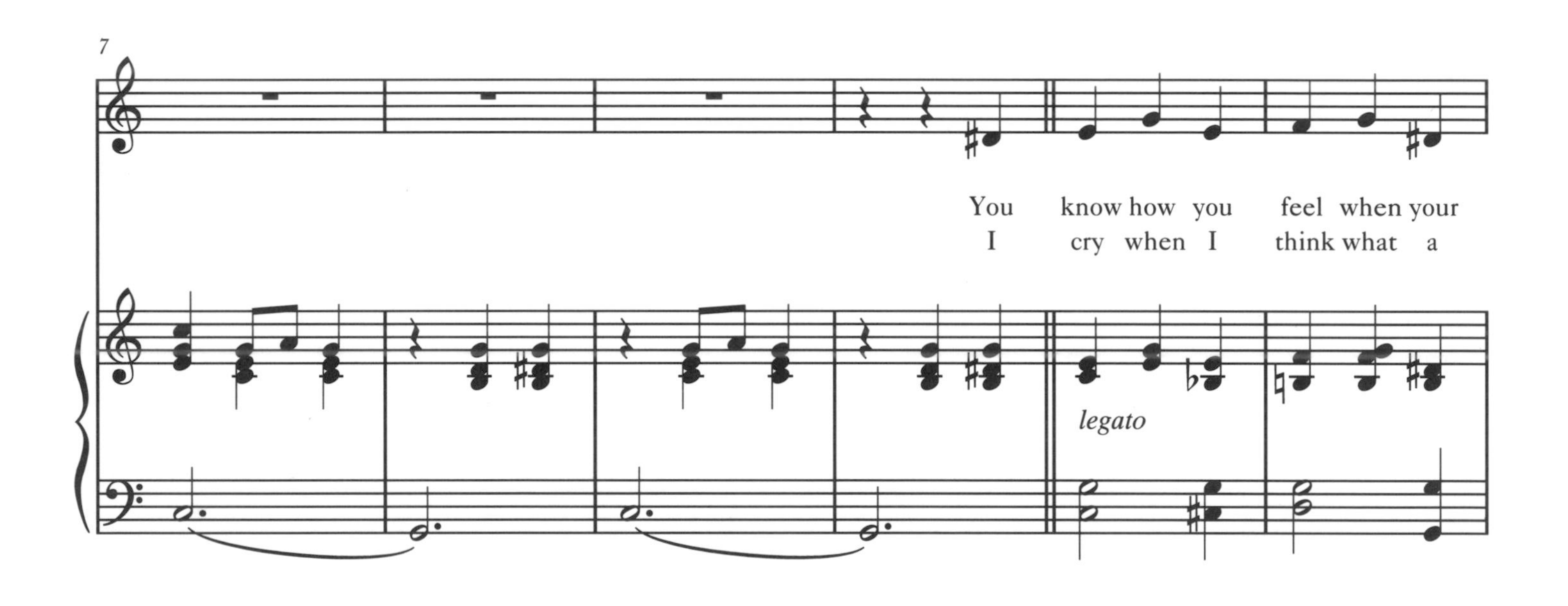

18
Well, that's how I felt when I first met Miss Le - na,
When she broke a glass why it would - n't be broke, Ach!
23
sweet lit - tle neat Le - na Kraus. I just hate to
no, it would on - ly be bent. To cry o - ver
28
think how she came in my life, Like a land - lord what comes for the
milk that's been spilt is - n't good, But the feel - ing that I'm feel - ing
33
rent, The rent was my heart when she took it a -
now, It was - n't just one glass of milk that I

38
part, Just how she came she went.
spilt When Le - na went, then went the cow.
43
CHORUS
Oh when I think of my Le - na,
My
I
p–f
47
face turns so white just like starch;
think of a girl who could cook,
The hand or - gans
Some sweet sau - er -
52
play pret - ty love songs all day But to me it's just plain fu - n'ral
kraut that would swim in your mouth, Like the fish - es that swim in the

57
march. It's bad to be sad, to be glad when you're
brook. When I think of a meal that was real I just
62
sad Can't be did as you all must a - gree, If you
feel, Like a ship with no sail on the sea, There's no
67
feel like I feel and you hear my ap - peal,
tears in my eyes, but my ap - pe - tite cries
Oh! bring back my
72
Le - na to me. me.
1.
2.

56. THAT KAZZATSKY DANCE

Words and Music by
IRVING BERLIN

13
A - bie take a chance. Take me by the wrist with your
hate you, love you so. Look at what you did, Oh you
16
fist, make a twist Or else I'll go right in a trance.
kid in the lid, I'm go - ing cra - zy like a loon.
19
Come and give your ba - by just one glance And we'll do that Yid-dish - er dance.
Make some spoon-ing with a sil - ver spoon, While they play that Yid-dish - a tune.
23
CHORUS
Oi, that Kaz - zat - sky dance, It makes me
p–f

26
lose my sense, come and get han - dy it's dan - dy, dan - dy,
30
Kiss me Kid, I'm can - dy. Oi, that Kaz - zat - sky dance, I'm go - ing
34
in a trance, I love my ham and cab - bage, kid, But Oi, that Yid-dish - er,
38
That Yid-dish - er, Oi, that Kaz-zat - sky dance. dance.
1.
2.

57. WISHING

Words by
IRVING BERLIN

Music by
TED SNYDER

9
Grant - ed by some fair - y brave;
When my ev - 'ry sigh means you?
11
Let's sup - pose a fair - y rose in fair - y clothes from head to toes, And
All the gold the world may hold its wealth un - told, I would un - fold, To
13
grant - ed us the wish we'd crave.
hold a place that just holds two.
15
Would you wish for dia - monds, dear, or would you wish for pearls,
Hold a place that holds a face, that holds a case of bliss,

17
Would you wish to rule the land and sea;
And that case of bliss would hold you too,
19
Would you wish a crown to rest up - on your gold - en curls? The
Such a place would make me trace up - on your face a kiss; The
21
maid - en sweet - ly an - swered "Not for me."
fel - low an - swered, "here's just what I'd do":
23
CHORUS
I'd wish for a night in June, A sil - v'ry moon real soon; A

27
moon that makes you want to spoon, And soft - ly croon love's
30
tune.
Then a tree that I could trust, A
33
bench that just holds two;
Then I'd wish for Cu - pid's
36
lov - ing dish, And then I'd wish for you.
1.
2.
you.
D.S.

58. DAT'S-A MY GAL

Words and Music by
IRVING BERLIN

10
pick and shov', Feel so gay __ that I say, __
bar - ber shop, Once he said __ when we wed, __
13
If I die to - mor - row, hon - est boss I no feel sor - row 'Cause I
If the bus - i - ness is rot - ten he's a shave - a me for not' - in'
15
feel so fine __ all the time Got - a such - a feel - ing like da
And he's got __ some gal what Man - i-cure my fin - gers ev - 'ry
18
cham - pagne wine. __ Since I met - a my
day for not'. __ Feel so glad __ that I

20
sweet An - nette - a much love I get - a for mine.
can't get mad and I feel yet bet - ter than that.
23 CHORUS
My gal, she's - a got - a such - a fig - ure,
p–f
25
So big, may - be it's a lit - tle big - ger; Small feet
28
just - a like - a Jap - a - neese - a, Big waist, it

30
take me sev - en days to squeeze 'er; My gal,
32
she's a dream - a, peach - a, cream - a, Sweet to beat the band and
35
When you see a miss who got - a shape - a like - a diss,
37
1.
2.
Dat's - a my gal.
gal.
fz

59. THAT DYING RAG

Words by
IRVING BERLIN

Music by
BERNIE ADLER

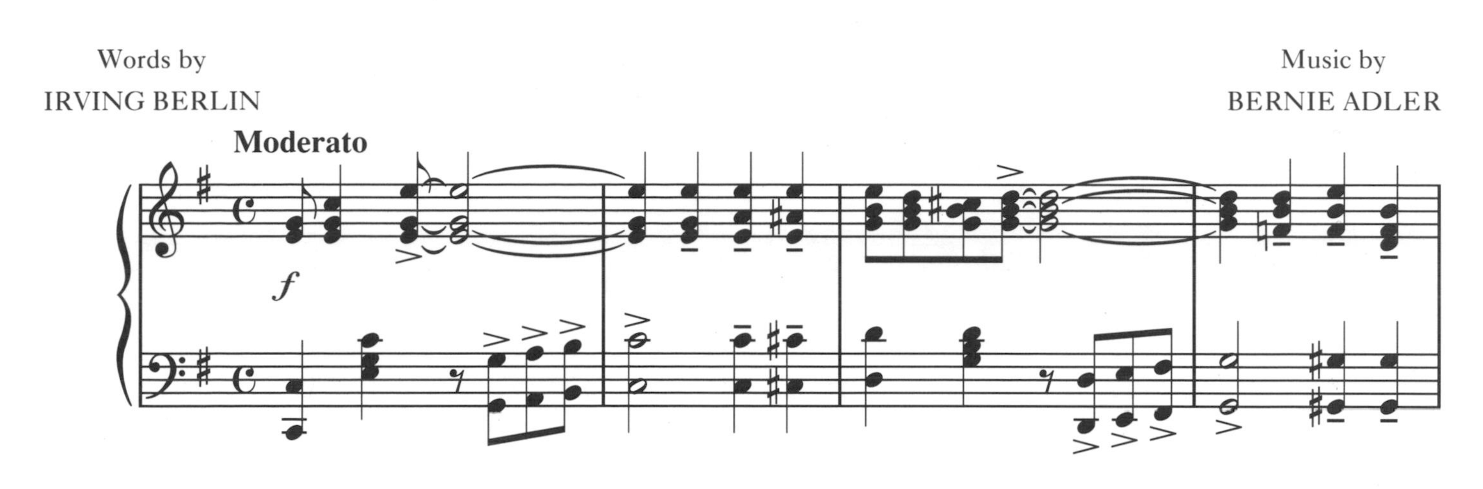

12
Soon I will breathe my last, Fe - ver is ver - y high,
Soon it will be too late, Heav - en - ly gates I see,
14
Doc - tor said I will die; 'Fore I take my long, long rest,
St. Pe - ter's call - ing me; Thank the Lord I'm gwine to go,
17
Grant me, dear, one re - quest. 'Mem - ber that af - ter - noon,
Up a - bove, not be - low. Right on my tomb - stone, dear,
20
We heard that dream - y tune, When you re - marked to me,
Put this in - scrip - tion clear, "Bur - ied with - out dis - grace,

22
Who wrote that mel - o - dy, That's the air, I
Died with a smil - ing face." Hon - ey say I
24
want to share be-fore I die.
passed a - way with-out a care.
CHORUS
27
p–f
Hon-ey dear, I must be go-ing I fear,
30
I feel the fin-ish is near, Please bring your

33
cel - lo in here,
While I see
36
just play the sweet mel - o - dy,
And hon - ey,
39
call it for me,
"That Dy - ing Rag."
1.
42
Rag."
2.

60. ALEXANDER'S RAGTIME BAND

Words and Music by
IRVING BERLIN

16
Oh, ma hon - ey,
Let me take you to Al - ex - an - der's
Come and lis - ten,
To a clas - si - cal band what's peach - es,
19
grand - stand, brass band,
Ain't you com - in' a - long?
come now, some - how,
Bet - ter hur - ry a - long.
22
CHORUS
Come on and hear,
Come on and hear
Al - ex - an - der's rag - time
p–f
26
band,
Come on and hear,
Come on and hear,
It's the best band in the

30
land, They can play a bu-gle call like you nev-er heard be-fore,
33
So nat-u-ral that you want to go to war; That's just the
36
best-est band what am, hon-ey lamb, Come on a-
39
long, Come on a-long, Let me take you by the

42
hand, Up to the man, Up to the man who's the lead - er of the
46
band, And if you care to hear the Swa - nee Riv - er played in
50
rag - time, Come on and hear, Come on and hear Al - ex -
53
an - der's rag - time band. Come on and band.
1.
2.

APPARATUS

SOURCES

Primary sources for the 190 songs comprising the body of this edition, and for the seven piano pieces included in the appendix as A18–A24, are the earliest sheet music versions deposited for copyright in Washington, D.C., and then marketed. These represent each song as Berlin wanted it to reach the public at the time of publication.

Once a song had been engraved, proofed, sent off for copyright, and printed in quantity, further changes were never made in the plates; subsequent printings, even years after first publication, gave precisely the same readings.

Some songs remaining popular over many years, however, were revised according to changing performance conventions or to make them accessible to a wider public. Ukulele or guitar tablature grids were often added, and some songs were arranged for a different performance medium such as keyboard solo, or voice with zither accompaniment. Folio collections of songs in simplified arrangements for performers with elementary keyboard and vocal skills were brought out periodically by Berlin's publishing house. Some early songs that remained popular into the 1930s or 1940s, or that were revived for a film of the period, were published in new arrangements that usually lacked the second verse and included a reharmonized or simplified accompaniment.[1]

Though such later versions of Berlin's early songs document changing styles, the inclusion of all of them would more than double the size of the present edition and would shift its focus away from the songs as they were known to performers and audiences before 1915. Therefore, no later revised versions of the songs are included.

Other materials have been collated with the primary sources. The Irving Berlin Collection in the Library of Congress (LC-IBC), though containing chiefly materials from later stages of Berlin's career, nevertheless has some valuable documentation relating to the early songs: typed and handwritten lyric sheets, a few lead sheets, lists of songs drawn up by Berlin himself, and scrapbooks of newspaper clippings. Where relevant, these materials are incorporated into the critical commentary. Also, many of these songs

[1]A new arrangement of "When I Lost You" copyrighted in 1939 not only simplifies the piano part and drops the key from C to B-flat, but even forces this durable waltz ballad into duple meter.

were recorded on cylinders or discs during the 1910s, sometimes with verses, second choruses, and catch lines not found in the sheet music. Such additional material has been added to the critical commentary.

The first sixteen songs in the Appendix were never published, thus other source materials were used. Exceptionally for the period, A1 and A2 were deposited for copyright as printed lead sheets—melodies with underlaid text, with no piano accompaniment. Manuscript piano-vocal scores must have existed at one point, but they have not been recovered. A3–A7 were recorded but never published. With the exception of a typed lyric sheet for A7 in LC-IBC, no written texts or music have been preserved for any of them. Their melodies are given here as transcribed from the recordings, with first and second endings of the chorus and such indications as D.S. and vamp specified to bring them into visual conformity with published songs. Since singers of the period tended to treat their material with considerable freedom, it was often difficult to distinguish between pitches and rhythms in their performances and those intended by Berlin. The accompaniment was often a help, since instrumentalists doubling the vocal line tended not to embellish the melody. Eight songs, A8–A15, were never copyrighted or published.[2] They are preserved in LC-IBC as textless lead sheets, with lyrics on separate sheets. This edition brings music and text together. Any problems in text underlay are reported in the critical notes. Lyrics and melody of A16 were printed in a British newspaper. Sections of the melody seem problematic, raising questions about the accuracy of the transcription, but there is no other source for the song and it appears here as printed.

The seventeenth song in the Appendix, "Father's Beard" (A17), was copyrighted and printed as a pamphlet. The first twenty-two stanzas are recited, the last is sung to the tune of "The Rosary," by Ethelbert Nevin and Robert Cameron Rogers. The first section of Nevin's song, as published by G. Schirmer in 1898, is given here, underlaid with the text of Berlin's last verse.

In line with the policies of this edition, it was decided not to add newly composed piano accompaniments to songs preserved only as melodies. More specific details concerning sources are given in the critical commentary for individual songs.

Editorial Method

Intended for both performance and study, this is a critical edition. Obvious mistakes in music and text have been corrected, consistency has been brought to certain minor details of musical notation and orthography, and period recordings and sketches have been collated against the primary sources.

The music is newly set, rather than being offered in facsimile, for two reasons: (1) the original plates, engraved by various craftsmen working from scores written down by a succession of music copyists, contain some mistakes in both music or text and many more inconsistencies in orthography and details of musical notation; and (2) though the original engraving was done by skilled professionals, their work varies considerably in layout and general appearance; thus, having everything newly set results in a visual consistency not possible in a facsimile edition.

So as not to distract the performer, the songs are given here free of brackets, asterisks, footnotes, and other visible signs of editorial emendation. Original readings for passages with editorial changes are recorded in the critical notes.

[2]A typed rubric at the bottom of each lyric sheet states that the song was entered for copyright, but a search of the copyright files at the Library of Congress revealed no evidence that the songs had been copyrighted.

More specific editorial procedures for dealing with mistakes, inconsistencies, questionable passages, archaic notational practices, and various other details are explained in what follows.

Song Titles and Attributions

Title and names of lyricist(s) and composer(s) are given as they appear on the first page of the music, not the cover, since the cover artists sometimes worked with incomplete or incorrect information. A title or attribution on the front cover that conflicts with the first page of music is reported in the critical commentary.

Rubrics and Repeats

The indications *piano* and *voice* found in some of these songs have been removed, for lack of significance and consistency.

The several repeated measures introducing the verse are sometimes labeled *vamp*, sometimes *till ready*, and occasionally both or neither are present. No attempt has been made to bring about consistency, for lack of significance. When either term is present but repeat signs are missing, the repeats are added and reported. No repeats are added when neither is present. A few songs have repeat signs in the voice part of the vamp even though the voice contains a pickup to the verse. Such repeats are misleading and therefore have been silently omitted.

The designation *refrain* is occasionally used rather than *chorus*. This difference has not been reconciled, for lack of significance, but where neither term appears, *chorus* has been added at the appropriate point.

A few songs, mostly high-class and romantic ballads, have a da capo (D.C.) indication at the end of the chorus to indicate a return to the piano introduction before the second verse. Various sets of signs used to indicate the more common dal segno (D.S.) return to the beginning of the vamp have been standardized as follows: D.S. is placed below the lower piano staff in the second ending of the chorus; the sign itself appears above the final double bar of both voice and accompaniment, and above the beginning of the vamp. Some songs have dal segno signs at the beginning of the vamp or at the double bar of the second ending, but not in both places. In such cases, the appropriate indication has been added silently. Where neither D.C. or D.S. is found, however, none has been added in this edition, leaving open the question of whether a return to the introduction or the vamp is intended.

Repeat signs at the first ending of the chorus were understood to indicate a repeat of the chorus rather than a return to the piano introduction or vamp at this point, thus repeat signs are not necessary at the beginning of the first bar of the chorus. Occasionally they are present, though, and have been retained because of possible significance, such as yet another repeat of the chorus.

Dynamic and Tempo Indications

The choruses of many songs are marked ***p* - *f***, implying a louder dynamic level for the second chorus than the first, but this indication is not added to those songs lacking it.

Dynamic markings are not added to the voice if they are found only in the accompaniment.

No metronome markings appear in the original sheet music. There are, rather, tempo markings, usually in Italian, retained in this edition. In keeping with the flexible performance practice of popular music, no metronome markings have been added in this edition.

A discussion of what may be learned about dynamics and tempo from period recordings of these songs is given in the introductory essay in part I of this edition.

Pitches

The pitch of a note is changed only in the case of obvious error, or if a parallel passage gives a better reading. All such alterations, which are not numerous, are reported in the critical notes. Where some detail of the original is not clearly incorrect but another reading seems preferable, the original is retained and an alternative is suggested in the critical notes.

Accidentals

Accidentals hold for an entire measure but no more. If an accidental governs the last note of a measure and the same note is repeated as the first note of the following measure, the latter note requires an accidental sharp, flat, or natural, to avoid confusion. Likewise, if a measure, or one voice within that measure, begins with a rest followed by a note repeating the last note of the previous measure, which note was governed by an accidental, this first note must have an accidental sharp, flat, or natural, for clarity.

Accidental signs are given as in the primary sources, even in cases where modern usage would prescribe an enharmonic spelling.

The primary sources reveal considerable disparity from song to song, and even within a single song, in the use of cautionary accidentals—sharps, flats, or naturals not strictly necessary but nevertheless given to prevent the performer from singing or playing the wrong note. A passage in which a cautionary accidental is used might be followed by a similar or parallel passage with no such accidental. One need look no further than "Alexander's Ragtime Band" to see the problem: in the third measure of the chorus there is no cautionary natural before the g′ on the second beat of the voice part, even though this note had been altered by a sharp only a beat earlier; but in the right hand of the piano in the fifteenth measure of the chorus a cautionary flat originally appeared before b′, even though this note is preceded in the same measure by a b-flat in the left hand, and its harmonic function is unambiguous.

In all likelihood the use of cautionary accidentals doesn't represent the intentions of Berlin himself so much as decisions made by a succession of scribes, arrangers, and engravers. Many of these accidentals are useful in alerting the performer to an ambiguous situation, and others serve to signal a modulation or tonal shift. They have been retained with the exception of those so redundant as to create confusion rather than clarification. Cautionary accidentals not found in the original sheet music have been added only in passages where their absence seems to risk wrong notes. All removed or added cautionary accidentals are reported in the critical notes.

Stemming

Note stems appear as in the original notation, except in the case of obvious error. No attempt has been made to add extra stems for piano passages in which contrapuntal identity of several lines is not intended. The now-obsolete double stemming of certain notes whereby one stem indicates that the note is to be sustained while a second stem in the opposite direction shows that the note is also part of an extended passage, has been retained except in cases where such stemming inhibits easy reading. In the voice part, double stemming has been introduced in a few places where two verses have different syllabification.

Articulation Marks

Where two or more successive chords (or octave passages) in the piano part are slurred, there is inconsistency in the original sources: sometimes a slur is used for each note of the chord, sometimes a single slur serves for the entire passage. This edition uses the modern convention of a single slur in all such cases to make a clear visual distinction between tied and slurred chords.

Berlin's text settings are almost exclusively syllabic. Where, exceptionally, two or more notes are sung to a single syllable, a slur usually appears over them. If such a slur is missing, it is added and its addition reported in the critical notes.

The greatest disparity in the notational practice of these pieces comes in the use and non-use of marks of articulation, such as slurs, staccato dots, and accent marks. Usage varies from song to song. Moreover, similar and even identical passages within the same piece may be articulated differently, or one passage may have articulation marks while a parallel passage has none. Any attempt to bring complete consistency to these matters would open up a bottomless pit of editorial intervention, and would represent a radical departure from the original sources.

It must be kept in mind that the notational system of popular music of this era was not intended to specify all details of performance. Singers and pianists were assumed to understand how the music should go, and to be able to decide for themselves how a given phrase should be articulated. It should be noted, however, that a passage is frequently given articulation marks on its first appearance, but these marks are missing in subsequent identical or similar passages; in other words, the performer is often shown how a passage should be articulated the first time it appears, but not after that.

Articulation marks have been added or changed sparingly in this edition, and then only in specific situations. In accordance with the principle suggested at the end of the previous paragraph, if a passage has no articulation marks when it first occurs but is supplied with such marks later in the song, these are added to its first appearance, and reported in the critical notes.

Where articulation slurs are found in the accompaniment but not in the voice, they have not been added to the latter.

Punctuation, Capitalization, and Spelling

Obvious errors in punctuation are corrected. Otherwise, original punctuation is retained, with the following five exceptions:

1. Periods appearing after song titles, names of authors, dynamic markings, tempo indications, and other rubrics are deleted, for lack of consistency and significance. However, in most songs, periods appear after the numbers indicating first and second endings, and these are added where lacking.

2. Where no punctuation appears at the end of a line of text which comes to a close, a period is added.

3. Where the sense of a phrase or line of text is compromised by the absence of punctuation, a comma, semicolon, or period is added, as appropriate.

4. Where a character speaks in the first person and quotation marks are found either at the beginning or the end of the passage, but not both, the missing quotation mark is added.

5. Quotation marks are not used when the text of an entire song is in first person. But in a few instances, the original sheet music does use quotation marks for a chorus in first person following a verse in third person. Since such cases are rare and conform to no discernible pattern, the quotation marks have been deleted, for consistency.

The prolongation of one syllable of text over the duration of two or more tied notes is sometimes indicated in the original by a dotted line, sometimes by an extended dash, and sometimes not indicated at all. A standard extender line appears throughout the present edition.

Most original spellings and capitalization are retained, even if they differ from today's usage. Obvious errors are corrected, however, and spelling has been modernized in a handful of instances where the original form might cause confusion.

All changes in punctuation and spelling are made silently.

Word Breaks

Where word division is lacking or conflicts with current usage, it has been brought into conformity with *Merriam-Webster's Collegiate Dictionary,* 10th ed.

Critical Commentary

Information for each song is reported in the following order: copyright number and date; publisher (or source, if unpublished); cover artist; alternate title on front cover; textual or musical material quoted in song; show(s) in which song appeared, theater, and date of opening; anthology in which song is published; miscellaneous information; and critical notes.

The critical notes indicate editorial changes and queries in abbreviated form. The measure in question is identified first, followed by the number of the note within this measure (notes and rests of any denomination count as one item each, but the second of two tied notes is not counted); V, P1, and P2 specify, respectively, the voice part, the right hand of the piano accompaniment, and the left hand; then the *original* source reading is given, with pitches designated according to the system C-c-c′-c″-c‴, in which middle C is c′. Thus the entry "11.4, V, no sharp before c′" reports that a sharp not found in the original has been added before the middle C falling on the fourth note of the eleventh measure of the vocal line.

Editorial emendations are recorded only in the critical notes and are not signaled in the edition itself, which gives only the corrected reading.

All theaters are in New York City unless otherwise indicated.

Abbreviations

Sources are identified by the following abbreviations:

ARB	David A. Jasen, ed. *"Alexander's Ragtime Band" and Other Favorite Song Hits, 1901–1911.* New York: Dover, 1987	
FSA	*Feldman's 18th Song Annual.* London: B. Feldman, 1912	
GY	*Irving Berlin's 90 Golden Years.* Ilford, England: International Music Publications, 1983	
IB	*The Songs of Irving Berlin.* New York: Irving Berlin Music Company, 1991 [six folios]	
	NS	*Novelty Songs*
	MS	*Movie Songs*
	BS	*Broadway Songs*
	PS	*Patriotic Songs*
	R&ES	*Ragtime & Early Songs*
	B	*Ballads*
LC-IBC	The Library of Congress; the Irving Berlin Collection (deposited in the Music Division, fall 1992)	
POMH	Stanley Applebaum, ed. *"Peg o' My Heart" and Other Favorite Song Hits, 1912 & 1913.* New York: Dover, 1989	
SIB	*The Songs of Irving Berlin.* Boca Raton, Fla.: Masters Music Publications, n.d. [seven volumes]	
TSLB	Sandy Marrone, ed. *"The St. Louis Blues" and Other Song Hits of 1914.* New York: Dover, 1990	

Months are given as follows: Jan, Feb, Mar, Apr, May, June, July, Aug, Sept, Oct, Nov, Dec.

1. MARIE FROM SUNNY ITALY

Copyright number and date. E150811; 12 Oct 1907
Publisher. Jos. W. Stern & Co.
Cover. Etherington
Miscellaneous. Large cover photo of Leah Russell
Critical notes. 5–6, P1, slurs probably indicate arpeggiation; 18.1, P2, unnecessary sharp before d′; 22, P1, no staccato

2. QUEENIE

Copyright number and date. E175522; 29 Feb 1908
Publisher. F. A. Mills
Cover. (a) Collective "house" cover listing titles of thirty-four songs published by Mills; (b) A. Ferraioli
Critical notes. 3.4, 36.4, and 44.4, P1, g′ not tied to first chord of next measure; 19.3, P1, natural before d′ rather than c′; 24.1, 35.1, 37.1, 45.1, and 47.1, P2, no accent; 38.3, V, eighth note

3. THE BEST OF FRIENDS MUST PART

Copyright number and date. E175845; 6 Feb 1908
Publisher. The Selig Music Pub. Co.
Cover. A. Ferraioli
Critical notes. 39.2, P2, unnecessary natural before g

4. I DIDN'T GO HOME AT ALL

Copyright number and date. E203451; 13 Mar 1909
Publisher. Maurice Shapiro
Cover. Starmer
Title on cover. I Didn't Go Home At All (I Heard the Clock Strike One A.M., One A.M., One A.M.!)
Anthology. SIB I
Miscellaneous. "As Introduced By Alice Lloyd, England's Dainty Chanteuse" (cover)
Critical notes. 2.6, P1, sharp before c″ rather than d″; 7.1 and 33.1, P2, no accent; 8.2 and 8.4, V, no flags; 12.4, P1, no sharp before c′

5. DORANDO

Copyright number and date. E203814; 11 Mar 1909
Publisher. Ted Snyder Co. (Inc.)
Cover. E. H. Pfeiffer
Anthology. SIB I
Critical notes. 3.3, P1, no quarter rest; 14.3, P1, eighth rest; 15.7, P1, eighth note e′ is dotted; 18.2, P1, no natural before g; 26.2, P2, a rather than g; 26.7, all staves, no eighth rest; 42.1, P1, no tie; 47.2, P2, B-flat should probably be A; 54.1, P2, "Cym." calls for a cymbal crash in orchestration

6. NO ONE COULD DO IT LIKE MY FATHER!

Copyright number and date. E204479; 2 Apr 1909
Publisher. Ted Snyder Co. Inc.
Cover. E. H. Pfeiffer
Title on cover. No One Could Do It Like My Father
Critical notes. 5.1, P1, sharp before f′ rather than g′; 25.2, P2, no natural; 42.1, P1, chord should probably be g/b-flat/d′; 44.1, P1 and P2, no staccato

7. SADIE SALOME (GO HOME)

Copyright number and date. E204481; 2 Apr 1909
Publisher. Ted Snyder Co. (Inc.)
Cover. E. H. Pfeiffer
Title on cover. Sadie Salome Go Home!
Anthology. SIB I
Critical notes. 2.3, P1, slur begins; 8.5, P1, redundant sharp; 32.2, P1, slur begins

8. MY WIFE'S GONE TO THE COUNTRY (HURRAH! HURRAH!)

Copyright number and date. E209740; 18 June 1909
Publisher. Ted Snyder Co. (Inc.)
Cover. Frew
Title on cover. My Wife's Gone to the Country Hurrah! Hurrah!
Anthology. POMH
Miscellaneous. Additional verses by Berlin were printed in the New York *Evening Journal*. "Oh! Where Is My Wife To-night?" (no. 15 in the present edition) is an answer song.
Critical notes. 25.4, 26.4, 33.4, and 34.4, P1, no eighth rest

9. JUST LIKE THE ROSE

Copyright number and date. E211289; 2 July 1909
Publisher. Harry Von Tilzer Music Pub. Co.
Cover. Etherington
Anthology. SIB I
Miscellaneous. Published in three keys: B-flat (low voice), D (medium), and F (high)
Critical notes. 29.1, P2, upper note is G; 31–32, P1, no tie for f-sharp; 32.1, P1, no sharp before a; 40.1, slur ends; 46.3, P1, unnecessary sharp before a′; 48.1, P1, no quarter rest

10. OH, WHAT I KNOW ABOUT YOU

Copyright number and date. E211292; 2 July 1909
Publisher. Harry Von Tilzer Music Pub. Co.
Cover. Etherington
Title on cover. Oh! What I Know about You
Miscellaneous. Written originally by Jos. H. McKeon, Harry M. Piano, and W. Raymond Walker; a letter from Walker to James Fuld, dated 23 November 1954, says, "I wrote the song with Joe McKeon and Harry Piani [*sic*] while playing at Kennedy's Cafe (formerly Kid McKoy's) and Harry Von Tilzer heard it there and liked the title and melody but not the lyric. Offered us $50 for it and said he would have a new lyric written to it but would put our names on as the writers when he published it. Berlin rewrote the lyric, maybe all of it or maybe part of it but I do know it was a great deal different from our original lyric."
Critical notes. 9–10, no repeat marks; 17.2, P1, no natural before c′; 18.3, P1, no rest, P2, two rests; 41.2, no natural before c′; 61.2–62.3, P2, no slur; 63.2, P2, natural before g rather than f; 71.1, P1, no natural before g′

11. SOMEONE'S WAITING FOR ME (WE'LL WAIT, WAIT, WAIT)

Copyright number and date. E211293; 2 July 1909
Publisher. Harry Von Tilzer Music Pub. Co.
Cover. (a) Gene Buck; (b) Gene Buck (same illustration, different title)
Title on cover. (a) Someone's Waiting for Me "We'll Wait Wait Wait"; (b) We'll Wait Wait Wait Someone's Waiting for Me

Quoted material. 26–28, second verse, phrase "I'm coming home soon" may be reference to Henry Clay Work's temperance song "Come Home, Father!" (1864)
Anthology. SIB I
Miscellaneous. "Words by Edgar Leslie, Music by Irving Berlin" (cover)
Critical notes. 5.2, P1, no flat before e′; 39.1, P1, no flat before a′; 41.1, no hyphen between ***f*** and ***p*** ; 69.2, P1, no flat before e′

12. DO YOUR DUTY DOCTOR! (OH, OH, OH, OH, DOCTOR)

Copyright number and date. E213051; 6 Aug 1909
Publisher. Ted Snyder Co. (Inc.)
Cover. Unattributed
Title on cover. Do Your Duty Doctor and Cure My Pain
Critical notes. 42.1 and 43.1, P1, no flat before e′

13. GOOD-BYE, GIRLIE, AND REMEMBER ME

Copyright number and date. E213052; 6 Aug 1909
Publisher. Ted Snyder Co. (Inc.)
Cover. (a) E. H. Pfeiffer; (b) Unattributed: large photos of Howard & Howard
Title on cover. (a) Good Bye Girlie and Remember Me; (b) Good-by Girlie and Remember Me
Miscellaneous. Music attributed to George W. Meyer on first page of music, but to Ted Snyder on cover (a)
Critical notes. 48.4, P1, no flat before g′; 58.4–5, P1 and P2, no staccato

14. WILD CHERRIES

Copyright number and date. E213300; 12 Aug 1909
Publisher. Ted Snyder Co. (Inc.)
Cover. E. H. Pfeiffer
Title on cover. Wild Cherries That Cooney Spooney Rag
Miscellaneous. First published as a piano rag by Ted Snyder (A21 in the present edition)
Critical notes. 24.1, P2, no accent; 30.3, P1, unnecessary natural before a′; 39.1, P2, no accent; 41.4, P2, no tie for b′-flat

15. OH! WHERE IS MY WIFE TO-NIGHT?

Copyright number and date. E214570; 16 Sept 1909
Publisher. Ted Snyder Co. (Inc.)
Cover. Frew
Title on cover. Oh! Where Is My Wife To-night
Quoted material. 67–70, "My Bonnie Lies over the Ocean" (1881), anonymous
Miscellaneous. Answer song to "My Wife's Gone to the Country" (no. 8 in the present edition)
Critical notes. 55.1, P1, natural before a′ rather than g′; 75.1–76.1, P2, no slur

16. SHE WAS A DEAR LITTLE GIRL

Copyright number and date. E216226; 5 Oct 1909
Publisher. Ted Snyder Co. (Inc.)
Cover. Frew
Show. *The Boys and Betty*, Wallack's Theatre, 2 Nov 1908
Miscellaneous. "Successfully Sung By Marie Cahill" (cover)
Critical notes. 2.5–6, 16.5–6, 18.5–6, P1, no slur; 8.5 and 10.5, V, no natural before g′; 19.2–4, P1 and P2, no crescendo; 31.5, V, no dot after b′-flat; 33.4, P1, no natural before f′

17. SOME LITTLE SOMETHING ABOUT YOU

Copyright number and date. E217166; 14 Oct 1909
Publisher. Ted Snyder Co. (Inc.)
Cover. Starmer
Quoted material. 17–18, phrase "Faint heart never won a lady fair" is an allusion to Cervantes's *Don Quixote*
Critical notes. 24.1, P1, no natural before c′

18. IF I THOUGHT YOU WOULDN'T TELL

Copyright number and date. E217168; 14 Oct 1909
Publisher. Ted Snyder Co. (Inc.)
Cover. E. H. Pfeiffer
Critical notes. 9.7, P1, unnecessary natural before e

19. I WISH THAT YOU WAS MY GAL, MOLLY

Copyright number and date. E217169; 14 Oct 1909
Publisher. Ted Snyder Co. (Inc.)
Cover. Frew
Critical notes. 3.4, P1, no natural before c′; 21.5–6, P2, no tenuto; 23.5, P1, no natural before a; 29.5, P2, no sharp before f

20. NEXT TO YOUR MOTHER, WHO DO YOU LOVE?

Copyright number and date. E218110; 26 Oct 1909
Publisher. Ted Snyder Co. (Inc.)
Cover. (a) E. H. Pfeiffer (two portraits, one male and one female, in separate decorative ovals); (b) Frew (couple seated, embracing)
Title on cover. Next to Your Mother Who Do You Love?
Miscellaneous. Published originally in A-flat, then reissued in G
Critical notes. 14.1, P1, no natural before c′; 29.2, P2, no ledger line for D

21. STOP THAT RAG (KEEP ON PLAYING, HONEY)

Copyright number and date. E219550; 24 Nov 1909
Publisher. Ted Snyder Co. Inc.
Cover. E. H. Pfeiffer
Title on cover. Stop That Rag Keep On Playing
Show. *The Jolly Bachelors,* Broadway Theatre, 6 Jan 1910
Miscellaneous. "Sung by Stella Mayhew" (cover)
Critical notes. 1.2, P2, bottom note of chord is c; 1.4–6 and 3.4–6, P2, no staccato; 5.2, P1, chord is f′/a′/c″ rather than d′/f′/a′; 5.1–3, 6.1–3, and 40.1–3, P2, no slur; 18.4, P2, no ledger line for lower note of octave; 28.4 and 36.4, P2, unnecessary flat before a; 42.1, P2, no ties to following chord; 44.3–5, P1 and P2, no tenuto

22. CHRISTMAS-TIME SEEMS YEARS AND YEARS AWAY

Copyright number and date. E219710; 2 Dec 1909
Publisher. Ted Snyder Co. (Inc.)
Cover. E. H. Pfeiffer
Title on cover. Christmas Time Seems Years and Years Away
Critical notes. 16.5, P2, G lacks flag; 22.1, 38.1, and 48.1, P1, no eighth rest; 23.2 and 39.2, P2, no eighth rest; 31.4, P1, no sharp before f′; 52.2–5 and 53.2–4, P2, no accents

23. YIDDLE, ON YOUR FIDDLE, PLAY SOME RAGTIME

Copyright number and date. E220516; 30 Nov 1909
Publisher. Ted Snyder Co. (Inc.); Feldman no. 437
Cover. E. H. Pfeiffer
Title on cover. Yiddle on Your Fiddle Play Some Rag Time
Anthology. SIB I
Critical notes. 35.1, P2, no accent

24. I JUST CAME BACK TO SAY GOOD BYE

Copyright number and date. E220879; 8 Dec 1909
Publisher. Ted Snyder Co. (Inc.)
Cover. E. H. Pfeiffer
Title on cover. I Just Come Back to Say Good-Bye
Anthology. SIB I
Critical notes. 11.1, P2, no stem on half note; 14.4, P2, no dot after g′

25. BEFORE I GO AND MARRY, I WILL HAVE A TALK WITH YOU

Copyright number and date. E221856; 31 Dec 1909
Publisher. Ted Snyder Co. (Inc.)
Cover. Frew
Title on cover. Before I Go and Marry I Will Have a Talk with You
Anthology. SIB I

26. THAT MESMERIZING MENDELSSOHN TUNE

Copyright number and date. E222371; 22 Dec 1909
Publisher. Ted Snyder Co. (Inc.); Feldman no. 441
Cover. Frew
Title on cover. That Mesmerizing Mendelssohn Tune. Mendelssohn Rag
Quoted material. 23–26, 31–34, from *Lied ohne Worte,* opus 62, no. 6 ("Frühlingslied")
Anthology. SIB I
Critical notes. 12.2, P1, no natural before d′; 16.2–3, P1 and P2, no diminuendo; 20.1–2, P1 and P2, diminuendo displaced from 19.3, P2 (cf. m. 11); 36.7, P1, unnecessary flat before a; 37.1, P1, no b; 39.1, P1, no ties to b-flat and g

27. SOMEONE JUST LIKE YOU

Copyright number and date. E222372; 22 Dec 1909
Publisher. Ted Snyder Co. (Inc.)
Cover. Frew
Title on cover. Someone Just Like You, Dear
Quoted material. Opening of chorus suggests "Love's Old, Sweet Song" by J. L. Molloy
Critical notes. 17.2, P2, no chord; 17.4, P2, no f; 18.1, P2, no sharp before c; 18.2, P2, no g (changes to 17.2–18.2, P2, made in accordance with parallel passage in mm. 9–10); 22.2, P1, no flat before b; 22.7, P1, no natural before d′; 23.3, 27.3, 31.3, and 35.3, V and P1, half note misaligned with third beat in P2

28. TELLING LIES

Copyright number and date. E223314; 14 Jan 1910
Publisher. Ted Snyder Co. (Inc.)
Cover. Unattributed, probably John Frew
Miscellaneous. Composer's name given as Henrietta Blanke-Belcher on cover

Critical notes. 3.2, P1, no natural before c′; 5.7, P1, redundant sharp before f′; 20.7, P1, unnecessary natural before d′

29. SWEET MARIE, MAKE-A RAG-A-TIME DANCE WID ME

Copyright number and date. E224516; 18 Jan 1910
Publisher. Ted Snyder Co. (Inc.)
Cover. Frew, collective for show
Title on cover. Sweet Marie. Make-a-Rag-a-Time-a-Dance with Me
Show. The Jolly Bachelors, Broadway Theatre, 6 Jan 1910
Miscellaneous. "As Introduced By Emma Carus" (cover); first page of music reads "Copyright 1909," though actual copyright date is 18 Jan 1910
Critical notes. 36.1, P1, chord has f′ rather than a′

30. IF THE MANAGERS ONLY THOUGHT THE SAME AS MOTHER

Copyright number and date. E224517; 18 Jan 1910
Publisher. Ted Snyder Co. (Inc.); Feldman no. 439
Cover. Frew, collective for show
Show. The Jolly Bachelors, Broadway Theatre, 6 Jan 1910
Miscellaneous. "As Introduced By Emma Carus" (cover)
Critical notes. 72.4, P1, c′/d′

31. OH HOW THAT GERMAN COULD LOVE

Copyright number and date. E226408; 15 Feb 1910
Publisher. Ted Snyder Co. (Inc.)
Cover. Frew
Title on cover. Oh, How That German Could Love
Show. The Girl and the Wizard, Casino Theatre, 27 Sept 1909
Miscellaneous. "As Sung By Sam Bernard" (cover)
Critical notes. 8.1, P2, no accent; 24.1, P1, no accent; 29.2, P1, middle note of chord is f′; 41.3–42.1, V, "someone" (Berlin sings "something" on a 1909 recording, Columbia A-804); 76.1–2, P2, no accents

32. WHEN YOU PLAY THAT PIANO, BILL!

Copyright number and date. E228389; 16 Mar 1910
Publisher. Ted Snyder Co. (Inc.)
Cover. Frew
Title on cover. When I Hear You Play That Piano, Bill
Miscellaneous. Cover shows a white pianist, although text implies a black protagonist
Critical notes. 33.4, P1, sharp before c′ rather than d′; 36.4, P1, chord incorrectly includes a′ (cf. m 28)

33. DRAGGY RAG

Copyright number and date. E229419; 13 Apr 1910
Publisher. Ted Snyder Co. (Inc.)
Cover. DeTakacs
Title on cover. Dat Draggy Rag
Anthology. SIB I
Critical notes. 10.5 and 18.5, P1, redundant sharp before f′; 11.1, P1, no natural before f′; 34.4, V and P1, no natural before g′

34. DEAR MAYME, I LOVE YOU!

Copyright number and date. E229542; 18 Apr 1910
Publisher. Ted Snyder Co. (Inc.)

Cover. Frew
Title on cover. Dear Mayme, I Love You
Critical notes. 28.1, P1, natural before g′ rather than f′; 34.3, P1, no sharp before f; 38.3 and 40.3, P1, no rests; 49.1, P1, unnecessary rest; 49.2, P1, no sharp before f′

35. GRIZZLY BEAR

Copyright number and date. E230452; 19 Apr 1910
Publisher. Ted Snyder Co. (Inc.); Feldman no. 453
Cover. Unattributed
Title on cover. The Dance of the Grizzley Bear [later corrected to "Grizzly"]
Show. *Ziegfeld Follies of 1910*, Jardin de Paree Theatre, 20 June 1910
Miscellaneous. Vocal version of "The Grizzly Bear Rag" by George Botsford (A22 in the present edition); song published successively with large cover photos of three singers: Maude Raymond, Tim McMahon, Sophie Tucker
Critical notes. 3 and 21, awkward cadences duplicate those found in the piano rag; 19.5, P1, no natural before a′

36. CALL ME UP SOME RAINY AFTERNOON

Copyright number and date. E230607; 23 Apr 1910
Publisher. Ted Snyder Co. (Inc.); Feldman no. 643
Cover. Frew
Anthology. SIB I
Miscellaneous. A recording made by Ada Jones in June of 1910 (Victor 16058-B) has the following second chorus:

Call me up some rainy afternoon,
Then again how's the evening for a spoon?
Call around tomorrow night,
We can then put out that fire in the furnace;
My Mama will sure be out of town,
She'll be entertained by Mr. Brown,
My Papa won't be 'round, he will call on Mrs. Brown
Angel pet, don't forget, goodbye.

37. THAT OPERA RAG

Copyright number and date. E230794; 28 Apr 1910
Publisher. Ted Snyder Co. (Inc.)
Cover. Frew, collective for show
Show. *Getting a Polish*, Wallack's Theatre, 7 Nov 1910 (originally titled *Mrs. Jim*)
Quoted material. 39–50, "Votre toast," from *Carmen* (1875) by Georges Bizet; 55–59, sextet from *Lucia di Lammermoor* (1835) by Gaetano Donizetti; 60–63, "Home, Sweet Home!" from *Clari* (1823) by John Howard Payne and Henry Bishop
Miscellaneous. "May Irwin's Song Successes" (cover); cover also lists "My Wife Bridget" and "He Sympathized With Me," though neither song was ever copyrighted or published
Critical notes. 8.1, P1, no ties from preceding chord; 9.3, V and P1, no ties to following note and chord; 14.1, P2, no sharps before octave G/g; 23.1, P1, unnecessary flat before b′

38. I'M A HAPPY MARRIED MAN

Copyright number and date. E231098; 2 May 1910
Publisher. Ted Snyder Co. (Inc.); Feldman no. 443
Cover. Etherington
Critical notes. 2.1–5.1 and 6.1, P1, no staccato or accents

39. I LOVE YOU MORE EACH DAY

Copyright number and date. E232638; 19 May 1910
Publisher. Ted Snyder Co. (Inc.)
Cover. Frew
Critical notes. 41, ***p*** *-**f*** dynamic indication uncharacteristically supplied above voice part as well as for piano; 54.1, an atypically weak progression; more characteristic would be A^7 (with a sharp before the c′ in P1) followed by D^7 (necessitating a natural before the c′); 72.3, V, quarter rest

40. ALEXANDER AND HIS CLARINET

Copyright number and date. E232639; 19 May 1910
Publisher. Ted Snyder Co. (Inc.)
Cover. Frew
Title on cover. Alexander and His Clarionet
Critical notes. 32.2, P2, no sharp before d

41. SWEET ITALIAN LOVE

Copyright number and date. E234395; 9 June 1910
Publisher. Ted Snyder Co. (Inc.)
Cover. (a) DeTakacs; (b) Frew, collective for show, large cover photo of Berlin
Show. *Up and Down Broadway,* Casino Theatre, 18 July 1910
Miscellaneous. Sung by Berlin in show; cover (b) omits Snyder's name, attributes song to Berlin alone
Critical notes. 36.2–37.1, P2, no crescendo; 37.1, P2, no staccato; 39.2, P1, and 39.3, P2, no staccato

42. OH, THAT BEAUTIFUL RAG

Copyright number and date. E235374; 7 July 1910
Publisher. Ted Snyder Co. (Inc.)
Cover. (a) Frew, for show *The Jolly Bachelors,* cover photo of Stella Mayhew; (b) Frew, for show *The Girl in the Kimono,* cover photo of Dale Fuller; (c) Frew, collective for show *Up and Down Broadway,* large cover photo of Berlin
Title on cover. That Beautiful Rag
Shows. (a) *The Jolly Bachelors,* Broadway Theatre, 3 Jan 1910; (b) *The Girl in the Kimono,* Ziegfeld Theatre (Chicago), 25 June 1910; (c) *Up and Down Broadway,* Casino Theatre, 18 July 1910
Miscellaneous. Sung by Berlin himself in *Up and Down Broadway;* originally published as a piano rag, by Ted Snyder (A23 in the present edition)
Critical notes. 3.1, P1, no natural before f′; 27.1, V, no ledger line for c′

43. TRY IT ON YOUR PIANO

Copyright number and date. E235475; 7 July 1910
Publisher. Ted Snyder Music Pub. Co.
Cover. DeTakacs
Show. *He Came from Milwaukee,* Casino Theatre, 20 Sept 1910
Anthology. SIB I
Critical notes. 2, no crescendo; 3.4 and 41.4, P1, no flat before b′; 4.2, P2, no accent; 15.4, V, no tie to following note; 33.2, P2, chord misaligned with 33.2, P1; 35.6, V and P1, no natural before d′

44. "THANK YOU, KIND SIR!" SAID SHE

Copyright number and date. E237664; 31 Aug 1910
Publisher. Ted Snyder Co. (Inc.)
Cover. Frew, collective for show

Title on cover. Thank You, Kind Sir!
Show. *Jumping Jupiter,* New York Theatre, 6 Mar 1911
Miscellaneous. Large cover photo of Richard Carle; cover attributes piece to "Berlin and Snyder"
Critical notes. 3.3, P2, no accent; 19.4, P1, eighth note misaligned with P2 sixteenth note (19.6, P2)

45. YIDDISHA EYES

Copyright number and date. E237828; 8 Sept 1910
Publisher. Ted Snyder Co. (Inc.)
Cover. Frew
Anthology. SIB I
Critical notes. 20.3, P1, no natural before a′; 30.7, P1, no flat before a′

46. IS THERE ANYTHING ELSE I CAN DO FOR YOU?

Copyright number and date. E237829; 8 Sept 1910
Publisher. Ted Snyder Co. (Inc.)
Cover. Frew
Title on cover. Is There Anything Else That I Can Do for You?

47. KISS ME MY HONEY, KISS ME

Copyright number and date. E238127; 3 Aug 1910
Publisher. Ted Snyder Co. (Inc.); Feldman no. 438
Cover. (a) Frew, large photo of Little Amy Butler; (b) Unattributed, for show *Up and Down Broadway,* large photo of Emma Carus; (c) Frew, collective for show *Jumping Jupiter*
Show. (a) *Up and Down Broadway,* Casino Theatre, 18 July 1910; (b) *Jumping Jupiter,* New York Theatre, 6 Mar 1911
Miscellaneous. Copyrighted on 11 Apr 1911 in an arrangement for voice and zither by G. Lechler (copyright no. E255489)
Critical notes. 19.4–6, P1, no staccato; 36.3–5, P1, rhythm given as two sixteenths and an eighth

48. COLORED ROMEO

Copyright number and date. E239224; 14 Sept 1910
Publisher. Ted Snyder Co. (Inc.)
Cover. Frew
Critical notes. 12.1, P1, no d′; 12.4, P1, no natural before b′; 13.2, P1, unnecessary natural before e′; 14.1, P1, no sharp before f′; 25.1 and 29.1, P1, unnecessary natural before a′; 27.3, P1, no ties to following chord; 34.1 and 42.1, f′ is not tied to preceding note; 41.2, P1, no d′

49. STOP, STOP, STOP (COME OVER AND LOVE ME SOME MORE)

Copyright number and date. E239367; 17 Sept 1910
Publisher. Ted Snyder Co. (Inc.); Feldman no. 478
Cover. Frew
Title on cover. Stop! Stop! Stop! Come Over, and Love Me Some More
Quoted material. 21–22, Berlin's "That Mesmerizing Mendelssohn Tune" (no. 26 in the present edition)
Anthology. IB:NS, SIB I
Critical notes. 2.1, P2, no accent; 2.3, P1, no natural before e′; 29.5, V and P1, no natural before g′; 32.3, P2, no quarter rest

50. HERMAN LET'S DANCE THAT BEAUTIFUL WALTZ

Copyright number and date. E240313; 24 Sept 1910
Publisher. Ted Snyder Co. (Inc.)
Cover. Unattributed, probably John Frew
Show. (a) *The Girl and the Drummer*, closed before opening, Aug 1910; (b) *Two Men and a Girl*, closed before opening, Dec 1910
Miscellaneous. "Successfully Introduced By Belle Gold" (cover); first eight bars of chorus are identical to the beginning of the chorus of "The Anniversary Waltz" (1941) by Al Dubin and Dave Franklin

51. PIANO MAN

Copyright number and date. E240646; 5 Oct 1910
Publisher. Ted Snyder Co. (Inc.)
Cover. Frew
Critical notes. 8.2, P2, lower note is f; 16.1–4, V, second verse text set under 16.1–3, with extender line under 16.4; 28.1, "L.H." appears, a measure late; 23.1, indication "2nd time 8va" appears over voice part but may have been meant for P1, since, at 42, pitch of notes implies that right hand should be played an octave higher for repeat of chorus; that solution, however, creates awkwardness in the crossed-hands passage (27–29) and in the high-register passage at the end of the chorus.

52. INNOCENT BESSIE BROWN

Copyright number and date. E240647; 5 Oct 1910
Publisher. Ted Snyder Co. (Inc.); Feldman no. 630
Cover. Frew (large photos of Mindell Kingston and Bessie Wynn)
Anthology. SIB I
Critical notes. 33.4, P2, chord is f/g/b

53. DREAMS, JUST DREAMS

Copyright number and date. E240951; 12 Oct 1910
Publisher. Ted Snyder Co. (Inc.); Feldman no. 446
Cover. Frew
Title on cover. Dreams Just Dreams
Miscellaneous. Published in three keys (C, E-flat, G), for low, medium, and high voices; edition for low voice, in C, has two errors in addition to ones corresponding to those described below: 11.2, P1, no sharp before d′; 35.4, P1, top note is d′
Critical notes. 15.3, V, sixteenth note; 17.2–4, P2, no slur; 26.2, P1, no natural before c′

54. I'M GOING ON A LONG VACATION

Copyright number and date. E240952; 12 Oct 1910
Publisher. Ted Snyder Co. (Inc.)
Cover. DeTakacs
Show. *Are You a Mason?*, closed during tryout, Apr 1910
Miscellaneous. "As Sung . . . By Miss Beth Tate" (cover)
Critical notes. 8.3–4, P2, no accents; 30.3, V, no accent; 32.1, P2, no ledger line for E-flat; 33.1 and 33.3, P1, no ledger lines for c′; 42.2, P2, lowest note is C; 45.1, P2, no ledger line for E-flat; 47.3, P2, no accent

55. BRING BACK MY LENA TO ME

Copyright number and date. E241527; 20 Oct 1910
Publisher. Ted Snyder Co. (Inc.)
Cover. DeTakacs

Show. He Came from Milwaukee, Casino Theatre, 20 Sept 1910
Miscellaneous. "Successfully Interpolated By Sam Bernard" (cover)
Critical notes. 34.2–3, P1, no quarter rests; 46.3, V, note is g′; 64.1, P1, no sharp before f′; 74.1, P1, chord is incomplete, an e′ should probably be added

56. THAT KAZZATSKY DANCE

Copyright number and date. E245819; 19 Dec 1910
Publisher. Ted Snyder Co. (Inc.)
Cover. DeTakacs
Critical notes. 1.2, P2, upper note is a; 2.3, P1, bottom note is c′; 3.3, P1, top note is d″; 4.2, P1, no staccato; 19.3, V, eighth note; 26.3 and 34.3, V and P1, d′ is probably an intentional dissonance to invoke "ethnic" sound; 36.3–4, P1, no accents; 39.5, P1, flat rather than natural before e′

57. WISHING

Copyright number and date. E246503; 17 Dec 1910
Publisher. Ted Snyder Co. (Inc.)
Cover. Frew
Show. (a) *The Girl and the Drummer,* closed pre-Broadway tryouts, Aug 1910; (b) *Two Men and a Girl,* closed during tryouts, Dec 1910
Miscellaneous. "Successfully Introduced By Elsie Ryan" (cover)
Critical notes. 16.6, P2, redundant natural before b; 16.7, P1, no quarter rest; 21.7, P1, no flat before e′; 26.4, P1, no quarter rest; 38.2–4, P2, passage is mistakenly written as C/c, D/d, E-flat/e-flat; 35, P2, no staccato

58. DAT'S-A MY GAL

Copyright number and date. E248729; 14 Jan 1911
Publisher. Ted Snyder Co. (Inc.)
Cover. (a) Frew; (b) A. B. Copeland (no illustration, simple ornamental design)
Anthology. SIB II

59. THAT DYING RAG

Copyright number and date. E250877; 18 Feb 1911
Publisher. Ted Snyder Co. Inc.
Cover. Frew
Title on cover. The Dying Rag
Critical notes. 5.4, P1, no ties to first chord in following measure; 13.1, P1, no natural before a′; 13.2, P1, no sharp before g′; 24.5, P1, no natural before c′; 29.4, P1, no accent

60. ALEXANDER'S RAGTIME BAND

Copyright number and date. E252990; 18 Mar 1911
Publisher. Ted Snyder Co. Inc.; Feldman no. 470
Cover. Frew
Show. Friar's Frolic of 1911, New Amsterdam Theatre, 28 May 1911
Quoted material. 47–50, "Old Folks At Home" (1851) by Stephen Foster
Anthology. ARB, FSA, GY, IB:R&ES, SIB II
Miscellaneous. "Successfully Intruduced [*sic*] by Emma Carus" (cover); also published as a march and twostep for piano (A18 in the present edition)
Critical notes. 5–6, no repeats; 37.5, P1, unnecessary flat before b′; 43.1 and 44.1, V, no ties; 50.3, P1, redundant natural before b′

SONG INDEX

Song titles are followed by their number in this edition (in parentheses), then the part in and page on which they appear. Well-known alternate titles appear in italics. Alternate titles that begin with the same word follow the entry for the title used in this edition. Those that begin with a different word appear as a separate entry.

Designed by Patrick Warczak, Jr.
Composed by A-R Editions Production Services
Set in Caslon 540 with oldstyle numerals
Music engraved by A-R Editions Music Production Services
using MusE, a program by Thomas Hall with Walter Burt